The Hidden Menace

The Hidden Menace

by

Maurice Griffiths

CONWAY MARITIME PRESS
GREENWICH

This book is a silent tribute

to all those officers and men of the Royal Navy, the Royal Naval Volunteer Reserve and the Royal Naval Reserve, and to their European, American and overseas Allies during two great wars, for their combined work in the laying, the sweeping, and the safe disposal of mines of all kinds.

First published in Great Britain 1981 by
Conway Maritime Press Ltd,
2 Nelson Road, Greenwich,
London SE10 9JB

ISBN 0 85177 186 6

Drawings by John Roberts
Typesetting by Computacomp (UK) Ltd, Fort William
Printed and bound in Great Britain by R J Acford, Chichester

CONTENTS

ILLUSTRATIONS

AUTHOR'S PREFACE

Many episodes in the long history of mine warfare cannot be included in this necessarily short review of the subject. It must be stressed, therefore, that this is not in any way an official account, but that it has been written more for the layman than for the expert. The selection of what has gone into this book, and the suggested possibilities of what the future might hold, together with any opinions expressed herein, are those of the author alone.

A book of this kind could not be written without extensive outside help. In this respect I should like to acknowledge my gratitude to Commander J G D Ouvry, DSO, RN (ret); Commander D A Irvine, DSC, RN (ret); the officers of the Instructional Minewarfare Section, HMS *Vernon*; the Naval Historical Branch of the Ministry of Defence; the Librarian, Britannia Royal Naval College, Dartmouth; and Mr Desmond Wettern, Naval Correspondent of *The Daily Telegraph*. All placed valuable documents at my disposal and afforded me unstinted assistance and advice.

Maurice Griffiths
Lt Cdr, GM, RNVR (ret)

PROLOGUE

There had been no formal declaration of war.

Shock waves echoed round the world when reports of the first sinkings were received. A big American freighter from New Orleans had been blown up in the Florida Channel some sixty miles off the Cuban coast, with half a dozen casualties. The remainder of her crew had managed to get away in the boats before their ship went down bows first.

The very next day, 21 August, the two-year-old missile cruiser *Canaveral*, after leaving her base at Norfolk, Virginia, on her way south, was sunk by a heavy explosion from beneath her when four miles off Cape Hatteras. No other vessel or submarine had been detected on the sonar, and the underwater upheaval which seemed to lift the ship bodily in the water struck as a complete surprise. This fine eight hundred million dollar warship was gone in less than half an hour, even though the watertight doors had been promptly closed, for the explosion must have ripped through most of her bottom plating. Happily, only a small number of her complement, mostly those who were below decks at the time, were lost. She was undoubtedly the victim of a deep laid mine.

From that day onwards the American public – hitherto engrossed in the Presidential election campaign – were to be shaken by the news of other ships being sunk by underwater explosions off the coasts of the United States. They included a 90,000-ton tanker in ballast as she was steaming through the Yucatan Channel on her way to Galveston, whose tanks blew up and tore her apart; a large freighter which was leaving San Francisco Bay for Yokohama; and a fully loaded container ship from Quebec for Liverpool as she was passing through the Cabot Strait off the Newfoundland coast.

The US Naval authorities acted with vigour, and groups of

specially equipped minehunters and Fleet sweepers were at once despatched to search and clear the danger areas. Until it could be certain that the main shipping lanes were clear, these channels were closed to all other traffic.

Once again, from the nature of the explosions which had wrecked all these ships it was almost certain that none of them had been torpedoed. An initial shock wave which had hit the vessel from underneath, like a battering ram, had been followed a second or so later by the main blast, which in each case seemed to spread the whole length of the ship, breaking her back, destroying her. All the signs pointed to a very heavy explosive charge either lying on the sea bed or moored many fathoms beneath the surface, and detonated by some influence which had been supplied by the ship.

Despite the array of new computers which had been installed by the Pentagon for the benefit of the World Wide Military Command Control System (casually known as WIMEX) the urgent question of what types of mines the Russians had been laying – how indeed they had got there – and exactly how they were armed and fired, still needed an answer. In London the Admiralty, although having access to the workings of WIMEX, was equally in the dark. For months past reports had been filtering through from agents behind the Iron Curtain, but they all lacked the information essential for the Allies to find the right antidote – how precisely the mine mechanisms were being actuated; by a ship's sound signature, by her magnetism in various ways, by variations of pressure in the water, or even by changes in light intensity beneath the hull.

'It's like 1939 all over again,' was a comment heard in the Operations Room in Whitehall. 'We've just got to get hold of a specimen or two to find out.'

But this time there were no young Luftwaffe pilots to lose their bearings and drop their precious cargo into shallow water where the British could later recover them. It was not quite like 1939, but the Admiralty as well as the United States Navy had formulated plans for demagnetising – the historic term was *degaussing* – all available ships. Once again, this was a gigantic task calling for thousands of miles of electric cables, hundreds of tons of equipment and multitudes of trained personnel to attend to all the principal ships of the NATO navies and the merchant fleets. It would take time to set into operation and carry out fully, and then the fact had to be considered: what if the Russians were not using magnetically armed or actuated mines at all, but combinations of other methods? The answer was sought only in the necessity to find out.

In British waters almost simultaneous casualties amongst shipping rocked the public and the Cabinet. Off the coast the first of a series of shipping casualties was the 10,600-ton merchantman, *Modern Commerce*, from Hull bound for Buenos Aires, which was blown up fifteen miles from Dover as she headed down-Channel. Her hull broke in the middle and the two halves sank in twenty minutes.

The following day another British-registered ship, the tanker *Golden Hope*, with a cargo of 80,000 tons of crude oil from the Persian Gulf bound for the Fawley refineries in Southampton Water, was racked by a deep explosion when about 25 miles north of the island of Guernsey. It would not be difficult to imagine the appalling conditions in her engine room when the first shock wave of the mine hit the hull beneath, rocking the boilers, lifting the turbines on their bedplates, breaking men's legs and backs. And then, perhaps one second later, the blast of the explosion bursting through her bottom plating, fracturing steam pipes, and in the sudden pitch darkness filling the cavernous engine rooms with scorching high pressure steam and the screams of men. And as the oil spilled out from her shattered tanks, set alight by the blast and spreading like a black carpet over the calm sea, it was a wonder that there were survivors who managed to get away in one of the boats.

These incidents were only the preliminaries to a number of sinkings of naval vessels as well as the bigger freighters which were to take place in waters around the coasts of Great Britain. The Royal Navy's sonar-equipped minehunters were already in action against the suspected ground mines, while to counteract those other mines, the ones moored in deep waters, stern trawler-type sweepers were ordered to operate over the suspected areas. Trained to hunt in pairs with their long sweep and its mine wire cutters set to be towed between them deep down near the sea bed, these useful units began to claim their first 'kills' when a mine rose to the surface with its severed mooring wire: but it blew up with a shattering roar, and still no specimen had been captured.

In addition to the naval minesweepers, aircraft of the RAF and USAF together with those of other NATO forces which had been fitted with elaborate electronic equipment designed to excite the hydrophonic type mine mechanisms were ordered into service. Together in European waters as well as those off the coasts of the United States, these valuable planes and helicopters carried out hundreds of sorties over suspected areas.

Within the first few days of the war other reports started coming in of ships being sunk, or in a few cases only badly damaged, by underwater explosions in other parts of the world. There were, for example, further casualties amongst the bigger merchant ships off the coast of California, in the Caribbean, off Savannah and off the coast of Newfoundland, but in addition came the news of a 10,000-ton refrigerator ship blown up in the Malacca Strait which, although badly damaged, managed to limp into Singapore. Another ship approaching the entrance to Port Jackson, New South Wales, was severely shaken by a nearby explosion which rose in a great mound of spray some 200yds away on her port beam, followed seconds later by a second unexplained explosion less than a mile away. It was reported later that both explosions had been far too heavy to have been a submarine's torpedoes.

With little doubt the most startling casualty of this kind occurred almost within sight of the American naval forces which had been keeping a regular patrol in the Arabian Sea area off the Gulf of Oman. One Russian nuclear submarine had already made its way into the Strait of Hormuz, dropped its cargo of mines in the main ship channels, and managed to slip away again, without being detected by the cruising American scanners.

In theory such an operation could not have been possible with the US underwater electronic eyes ever watchful, and the aircraft from the American carrier in the vicinity constantly on patrol; yet in war as in peace an unexpectedly daring move can sometimes escape the notice of the watchers. In this case the Russian commander approached his goal at slow speed, and at a great depth – as much as 450 fathoms – at which only nuclear submarines were able to operate. This, he knew, was too deep for the underwater detection devices the Americans were then using, and he kept his boat's speed down to 10kts, when it was possible to run in with virtual sonar silence. Only if he had to make a rapid getaway – and once detected he would have little chance of doing that even at his depth – would he risk working up to the 40kts his fine boat was capable of when submerged.

In the event the ploy paid off, and the submarine was able to return to its base to report mission accomplished. When the appointed day arrived for the mines to be rendered active, one of a stream of supertankers carrying their cargo of oil to Japan was blown up and set on fire fore and aft almost in the middle of the Strait. The channels were immediately closed to shipping, and the supply of oil from the Persian Gulf ceased from that day ...

* * * * *

A target which had long received the attention of the Kremlin was the group of oil and natural gas installations spread over the North Sea, and which had for some years now fed Great Britain and Norway with vital power supplies. It was clear to everyone that it would be all too easy to demolish, or at least seriously damage, the rigs themselves by rocket or concerted air attacks.

But the Politburo argued that as it was their intention to occupy England and Scotland as a priority aim during the earliest wave of attacks towards the North Sea coast, the total destruction of these elaborate and highly costly installations would later merely prove a regrettable loss of supplies to the Russians. It would obviously be to their advantage to capture, if possible, all the North Sea rigs in working order.

A much simpler and equally effective way of cutting off Britain's home oil supplies – the gas was not so vital – would be to blow up the pipelines themselves which carried the oil to the shore. The position of each of these pipelines was shown on charts specially prepared over the previous few years by Soviet spy trawlers, and at first the Politburo assumed that the way to destroy the eighteen principal pipelines was by delayed action mines which would be dropped in position from freighters on their apparently harmless voyages to and from British ports.

The reaction to this plan by the Soviet naval authorities was forcefully expressed as the sort of lunatic idea only a committee of landlubbers could conjure up: the North Sea, the admirals reminded their comrades, was not like a flat board with the exact lay of the pipelines marked in dotted lines across it. Any surface ship would have great difficulty in fixing the exact location, and if the mines were not dropped to within a few yards of the pipes damage could not be guaranteed. To be carried out in the greatest secrecy, they declared, the operation would have to be the work of minelaying submarines.

In the event, five diesel-electric submarines, three from Kronstadt in the Gulf of Finland, and two from the Kola Inlet, left on what was called Operation 'Artery', and made their approach undetected through the deep water off the coast of Norway. By means of their sensor detector devices they were able to locate each of their steel pipelines in turn, and lay an 800kg ground mine close beside the pipe. The mines were programmed to explode at some date between 21 and 24 August.

As the submarines left their respective sites each dropped in addition two magnetic/acoustic mines which had been set to respond to a small vessel. When the first underwater explosions occurred on 21 August naval minesweepers were despatched from Rosyth to investigate. One of the sweepers was sunk by a mine, while during the next two days ten more explosions caused the immediate shut down of the oil rigs affected ...

* * * * *

In Europe the massed land and air forces of the USSR had already begun their long-rehearsed offensive across the Iron Curtain line into West Germany towards Denmark, Holland and Belgium. The mighty westward thrust was met by the combined NATO armies and air forces, which were steadily forced to give ground as the Soviet tanks and armour pressed forward. For the third time within a space of seventy years, middle Europe was ablaze with the savagery of modern warfare.

From across the Atlantic the first of the heavy transport planes, which were later to be organised into a constant shuttle service, a mighty bridge of troop carriers spanning America to military bases in Western Europe, were assembling for take-off. This time America was concerned deeply in the war in Europe from the very beginning, and was already mobilising her vast resources.

Three years previous to these events the American public had been shaken into some awareness of the possibility – it had not yet seemed inevitable – of war with Russia, and an Act inaugurating a new form of draft, or military conscription, for women as well as men, had been passed through Congress. In this Great Britain still lagged behind, but not for much longer: mobilisation and conscription became the pressing need of the day.

On both sides of the Atlantic the NATO Powers were alert to the immediate need to mobilise more shipping for the transport of heavy war supplies as well as men, and to be ready to face a more intensive Battle of the Atlantic. For the Soviets their preliminary mining offensive seemed to have got off to a good start. In review, they had succeeded in laying their mines in secret, and in areas where they could cause the maximum of damage to the West's shipping, and predictably create some panic amongst the world's shipping lines.

The Soviet technicians, too, had long studied the minelaying policies of both the Allies and the Germans during World War II, and had introduced their own ingenious devices for evading the West's sweeping methods. They had reason to be pleased with their SK9 mine, which was designed to be moored in very deep water, while its very heavy explosive charge of 1450kg could sink a large ship from a depth as much as 90m beneath her.

Then there was the torpedo-mine, the PK14, designed for the shallower roadsteads and shipping channels, which rose from the bottom on its own power and homed-in on the victim with its 850kg charge. They already had variations of this mobile device which could be actuated in a number of ways. Fitted with their sensitive programming unit, all the mines were able to lie in wait while one vessel after another might pass by without exciting the mechanism, but immediately the ship with the right kind of sonic effects approached, the mine would, as it were, wake up and detonate. In addition to these devices the Russians were holding one other of their inventive arsenal of underwater mines, a secret weapon with such destructive power that it would not be brought into use without direct orders from Moscow.

The introduction of nuclear attacks on land targets was not in the Kremlin's programme. It was only too clear that the immediate response from the West would be retaliation in kind, which could only spell the essential destruction of the greater part of Europe and the industrial USSR. Nevertheless events were about to occur which would bring the outbreak of war very close indeed to the public. Glued to their television screens and morning papers as most people were, they were still obsessed with the news of the damage done to a number of the North Sea oil pipelines, and the effect this would be bound to have on the country's supplies.

The authorities had already agreed that there was nothing to be gained by attempting to suppress news of this kind, since they had learnt during World War II how well the British public usually reacted to bad war news: the famous 'blood, sweat and tears' outlook had merely stiffened their resolve to see the war through. The Russians in any case already had received full information on the success of Operation 'Artery' by means of their constantly orbiting satellites – as indeed had the Americans through their own watchful system.

Now, the next day, already feeling that there was a Red under every bed, the public learnt of the cataclysm that had overtaken the Bristol Channel. As more news came flooding in it appeared that people in Cardigan and Glamorgan, and in Devon and Somerset and elsewhere had experienced four distinct and very severe ground shakes about midday on the 27th.

'Like an earthquake, it was,' said one onlooker. 'It shook everything round me, and nearly knocked me over. And about a minute after that there was this deep rumbling sound, which seemed to have four distinct bumps, lasting perhaps another half minute. It sounded as if it was all out at sea somewhere.'

The day was bright and sunny with a light breeze out of the north west, yet despite the sunshine people living nearer the coast said they saw four distinct and intensely bright flashes of light on the horizon. 'As bright as lightning flashes on a dark night,' said one. 'They left me almost blinded for a time.'

They were not to know yet that Lundy Island in the entrance to the Bristol Channel had vanished for ever, nor that a few miles away a big, deep laden tanker, which was heading up for the oil terminal at Milford Haven, was overwhelmed, and disappeared beneath a gigantic wall of water. A number of other ships, two heading in, the others outward bound in the same area, encountered the same unbelievable tidal wave, and were never seen again.

Far out at sea four mushroom-shaped clouds rose into the air above the horizon, gradually mingling as they mounted high into the blue sky, reflecting from their midst all shades of yellow and grey and brown with the midday sun's rays on them. Gently, like a many-tinted storm cloud, the mass of vapour spread across the horizon until it had covered the sun, and began to drift slowly before the wind in a south easterly direction. It brought the cloud over the north coast of Somerset and part of Devon, and beneath it was gently falling a hardly visible curtain of white ash – a deadly fallout.

On the northern coast of the Bristol Channel a steady roaring sound grew in volume as the front of the enormous wave broke over the cliffs of Pembroke and thundered into the mouth of Milford Haven. Within the confines of this beautiful inlet the mass of water rose higher and higher and burst with a deafening roar almost simultaneously over the tanker jetties and oil storage tanks at Milford as it swept remorselessly over the quays and harbour installations of Pembroke Docks.

When the following surges, each a little less violent than the initial wave, had thundered into spray in turn and finally subsided into a sea of angry waters, there was little to be seen of the great oil terminal but wreckage, and the rusty bottom of an upturned tanker. And as the great mound of water with its tumbling white crest continued on its way up the Bristol Channel, it swept into Swansea Bay, drowning under a turmoil of water the docks at Swansea, Neath and Port Talbot in their turn.

Along the south shore the wave roared into Ilfracombe, Lynton and Minehead, destroying streets of houses and drowning thousands in their homes. And as the moving mountain of water began to press through the narrows between Barry and Weston-super-Mare, here only eight miles wide, eyewitnesses guessed its height at over a hundred feet. The two rocky islets of Flat Holm and Steep Holm in the middle of the channel disappeared, while in their turn Penarth, Cardiff Docks and the quays at Newport were overwhelmed in the surge.

'I hope I never see anything like it again,' a farm worker, who had been driving his tractor over a field on high ground several miles away, told a reporter. 'First I heard this sound like distant thunder. It grew louder every minute. Then I saw across the bay this bloody great mound of water coming up channel. It looked all dark green with a long white top that seemed to be falling forward all the time, coming along really fast. You know, like a train it was. And it just thundered over the shore down there and before you could say anything Clevedon, and all the boarding houses and hotels there, were gone, just buried under the sea.'

While urgent warning messages were being sent out to places farther up the Severn, there was nothing mortal man could do to arrest this thundering mass of water which was advancing at an estimated rate of fifty miles an hour – and gaining as it reached the narrower parts of the river. Many people could not believe the garbled reports as they came in from survivors farther down the coast. Those who lived near the Severn were only too familiar with the famous Bore, which came regularly up the river on the first of every high spring tide, a surge or wave rising in a series of steps to a height of four or five feet. It rushed between the banks at perhaps fifteen miles an hour, until it had spent itself in the narrow reaches above Blakeney, and if small boats took the Bore bows on they came to no harm. But the Severn Bore was nothing like this.

When the distant roaring sound came closer and the watchers on the banks by Portishead first witnessed the great cliff-like face of the wave advancing up river, they stood rooted to the spot, not quite believing what they saw. Part of the wave fell with a thunderous roar over the docks at Avonmouth, while its fury carried the surge headlong into the Avon, where it rushed up between the steep sides of the Gorge and beneath the graceful suspension bridge at Clifton.

By the time it had reached the lower docks at Bristol much of the relentless fury had been lost in the bends of the narrow river, and the damage done to the harbour buildings and dock installations was only minor in comparison with the devastation created lower down the estuary. Only the lower streets of the city were inundated, cars and buses swept along by the waters and people drowned in their basements: Bristol had been largely saved from destruction by its winding river.

As the remnants of the main surge continued up the Severn the land on each side of the river above Chepstow was flooded to a depth of ten or twelve feet, drowning livestock on farms in the Eastington area, before the rush of water welled along the banks towards the upper reaches of Gloucester. Here it seemed to have spent its strength, and paused before beginning its flow back towards the sea.

The amount of destruction the giant waves had caused to ships and port installations and buildings at Cardiff, Penarth, Barry, Port Talbot, Swansea, Pembroke and the oil terminal at Milford Haven was too extensive to comprehend at first. How many thousands had been killed by falling masonry or drowned in the flooded towns could not yet even be guessed at; nor could the long term effect of the radio-active fallout, which was still drifting over the countryside as the north-westerly breeze carried it across Devon and part of Somerset. This was a disaster of unexpected magnitude, and it left the people stunned.

The country had scarcely recovered from this catastrophe, and borne the grave news from Europe of the Russian armies' continuing advance across West Germany towards the Netherlands and the banks of the Rhine, when the second group of nuclear mines was exploded in the North Sea. If the effects of the four mines in the West Country had been disastrous, the results of these five were even more far-reaching.

The upheaval was timed for 1420 hours in the afternoon when a spring tide was already flowing towards the estuary of the Thames, and due to be high water at London Bridge at 1545. The mines were detonated at roughly 20-second intervals over a period of two minutes, and people who happened at the time to be on the cliffs between Southwold and Felixstowe, or on the coast from the Hook of Holland to Ostende, were blinded in varying degrees by the intensity of the five flashes. The earth shakes which followed immediately on the explosions damaged buildings in both the Dutch and Belgian coastal resorts, as well as in towns along the Suffolk and Essex coasts.

Minutes later came the blast waves, one after the other, carrying away roof tops and chimneys, breaking in doors and windows, and hurling people off their feet. The Shipwash lightvessel, together with the Outer Gabbard and the Noord Hinder on their stations which ran in a rough line between Orfordness and the mouth of the Scheldt, disappeared. Moments later the men aboard other lightships further south – the Sunk, the Galloper, the West Hinder, the Kentish Knock – stood helpless and uncomprehending as the huge mound of water reared up towards them, and overwhelmed their vessels one after another.

While the north-spreading waves roared into Lowestoft and Great Yarmouth harbours, and overwhelmed the sea dykes of Walcheren, Schouwen, Goeree, the Hague and Ijmuiden, eventually to spend themselves out in the North Sea, the frontal surge advancing south mounted even higher as it swept towards the narrows of the Dover Straits, roaring over the harbour walls of the historic Cinque Port, enveloping the ferry terminal and the Granville Dock. Along the coast the railway pier at Folkestone was swept away, while Ostende, Dunkirk, Calais and even Boulogne were overwhelmed in their turn as the series of waves passed down Channel gradually losing their destructive might.

Into the mouth of the Thames the waves thundered unchecked, rearing up to a height of 100ft and more as the Kent and Essex shores began to close in on them. The Great Flood of February 1953, when a fierce northerly gale in the North Sea drove a high spring tide southward before it, and the tidal surge rose some 15ft higher than normal, wrecking seaside homes and drowning hundreds of people asleep in their beds, was as nothing to this. The low-lying land along the Kent and Sheppey shores, the oil refineries on the Isle of Grain, and Canvey with its oil terminal and storage tanks, disappeared beneath the mountain of water.

So it went as the implacable waves thundered their way up London's river, drowning Gravesend and Erith on the south bank and Tilbury docks, Grays and Greenhithe to the north. The great Ford motor works with its loading quays at Dagenham was next as the buildings collapsed in the turmoil of foam.

Some four miles farther on the Thames flood barrier stood across the river by Woolwich. It had been designed and built to prevent a repetition of that disastrous 1953 flood, and was scheduled to be completed by mid-1978. But disputes over pay and working conditions and industrial action had created such delays in construction that the barrier had not been ready for the opening ceremony until February 1983. It was now all set to hold back any phenomenal high tide that was likely to flow in from the North Sea, and when the desperate warnings of the advancing waves were received, the routine drill was carried out and the openings in the channel were closed by the massive semi-circular gates. All ship movements were held up.

When the advancing wall of water with its great breaking crest a mile wide was first sighted by the machinery operators they could tell that it was far higher than their steel barrier.

'My God,' cried one, 'we'll all be drowned!'

It was to be the last thing he said as the great barrage disappeared and the water thundered on its way, to envelop all the streets of the City. There had been insufficient warning time to have all Underground trains stopped in stations, to cut off the current, and to get all passengers and staff up to street level. There had been no time even to close the flood doors at any of the stations on the Northern, Piccadilly, Victoria or Bakerloo tube lines while trains were still running.

The river rose 50ft in 15 minutes, and London's Underground system became flooded as far out as Clapham and New Cross in the south to Stratford, Camden Town and Baker Street to the north of the river. Inside a few days nine carefully laid five-megaton nuclear mines had created as much havoc and destruction, as well as loss of life, as had all the bombing raids over these areas during the previous war of 1939 to 1945.

Much of Central London was a drowned city ...

* * * * *

In pursuing their plan to occupy England and at least the industrial areas of Scotland as soon as the Red armies had swept across Europe to the North Sea ports, the Russians decided not to attempt to erase the principal cities in Britain (the phrase echoed an historic threat of Hitler's) by massive bombing raids or rocket attacks. The hideous destruction by bombing of the main centres of history and culture, which had been such a regrettable part of World War II in Europe, as well as in the United Kingdom, had shown itself in the end of little value to either side.

Well aware of this, the Soviets wanted Britain as an island fortress for their rocket bases, from which in due time they would be able to control all Europe and the whole of the North American continent as well. England and Scotland, not to mention Wales, with their industries and services in working order would be of far greater use to the USSR if captured intact (apart from inevitable demolition from within) instead of cities, factories, engineering works and shipyards wrecked by high explosives and fire. The British themselves would be dealt with as seemed expedient: forced to work for their new masters, or sent to labour camps in the USSR.

It was planned, therefore, to follow up the disruption created by the Severn and Thames mines and destruction of the North Sea oil pipelines almost immediately with Operation 'Aquarius'. This was the code name for a series of raids by high altitude aircraft whose crews had been trained to shower hundreds of small bombs on to all the principal reservoirs and waterworks throughout the country. With most of the main water supplies thus contaminated with nerve destroying chemicals, the greater part of the populace would become affected within a few weeks, and those who escaped would be in no condition to resist the invaders.

In the meantime, the Soviet plans to advance across Western Germany into Denmark, the Netherlands and Northern Norway, appeared to be going well in accordance with the schedule long before agreed upon in the Kremlin. The forces of the Western Alliance were steadily being compelled to give ground. But overseas, and in widely separated parts of the world, engagements were taking place between units of the United States, British and other NATO fleets and ships and submarines of the Soviet Union.

And the Russians were not having it all their own way. Although their navy in 1984 had greatly outgrown in numbers and strength of armaments those of America and Great Britain, when it came to monitoring an enemy vessel, interception, and engagement with the homing missiles employed, the technical training of the operators, the advantage of having free enterprise production in microelectronic equipment, and above all, the general superiority of American and British and other NATO naval training, proved in the end very much to the Allied forces' advantage.

On the other hand the Russians were holding in reserve a weapon which they had every reason to believe would eliminate any warship group, or better still, perhaps an entire convoy. And such a convoy was even now being assembled off the coast of New Hampshire, bound across the Atlantic with vital stores for the war zone in Europe. It was this large convoy that the Kremlin decided should be intercepted, and put to the test ...

* * * * *

When eastbound Convoy TE17 formed up in the region of Cape Cod, it was accepted that there would be little likelihood of heavy air attacks for at least the first three days of the voyage. The composition of forty-one merchantmen was accordingly ordered to maintain close formation of five lines abreast of eight ships each, the odd one, a 9000-ton freighter, bringing up the rear. As they steamed steadily eastward the ships formed a rough rectangle just over three miles in length and a mile across, with their naval escorts, including a small aircraft carrier, deployed around them like sheepdogs.

In this first part of their voyage all the ships' captains knew that the chief threat from the enemy awaiting them would be SLBMs (submarine-launched ballistic missiles) and long range homing torpedoes, also launched from submarines. It was in fact while this convoy was under attack three days later from both diesel-electric and nuclear submarines, that the Russians produced their carefully guarded secret.

The movements and changes of course of the convoy had been monitored and checked by computer, and while the missile and torpedo submarines were being heavily engaged by the group escorts – not without success with two submarines destroyed on the first day – the submarine *N265* passed across the estimated course some three miles ahead of the convoy at the maximum depth at which this class of nuclear submarine could operate. At a predetermined moment her commander gave the order, and a heavy mine of a special type was released. He then ordered full speed, and the boat raced off in an easterly direction at 40kts.

The mine, suitably programmed before being shipped aboard the *N265*, began to sink slowly in the silent depths, then its depth setting mechanism took over, and the great sinister object started to rise gently towards the surface, until it had reached a level of about 25 fathoms. Here the motor cut out and the mine remained close to this depth, rising and falling slowly in a world tinged a dark green from the sunshine above.

Hovering in this position, the mine waited while the great convoy of ships and escorts continued to steam steadily in this direction. When the first of the five lines of merchantmen were within perhaps half a mile, the mine sensor reacted to their combined sonic disturbance, and after a little delay, detonation followed.

As the huge mound of spray burst from the sea the three nearest ships were lifted up by the blast as though they had been merely toy boats. Their near neighbours also disappeared from view as a great fireball spread out on all sides. Ships farther down the lines were instantly enveloped, their superstructures melting in the unbelievable heat, the upperworks of others blistering and catching fire.

A vast mushroom of smoke and gases, shot through with livid flashes, rose above what was left of the convoy like a writhing storm cloud. Before the escort commanding officer at the rear of the columns could signal all remaining ships to alter course 90 degrees to the south'ard on full helm, many of the surviving vessels were already steaming into a fine curtain of spray and white ash which blotted out the sea in their path.

Only the last few ships took in the situation in time, and managed to turn away to starboard as smartly as big deep-laden freighters could do so. The deadly radio-active fallout from the nuclear fission mine was spreading gently over all the other units of the convoy which were still afloat. Each had become a death trap, impossible to unload, or even to handle in any way, in port. One mine had brought about the virtual extinction of a large convoy ...

Chapter one

UNDERWATER GROPING

If the definition of a mine as used in naval warfare is a device designed to explode against a ship so as to destroy her, the earliest recorded attempts to bring this about used not mines but contrivances that could best be described as floating bombs or petards. It could be argued that the Dutch were pioneers of this strategy, when at Antwerp in 1585 they drifted boats filled with casks of gunpowder against Spanish ships in the harbour, and exploded them with flintlocks actuated by clockwork. Compared with the comparatively small explosive charges of later sea mines, the Dutch operation could be described as being on the grand scale, and some hundreds of astonished Spaniards are said to have died in the attack, to the delight of the oppressed Hollanders.

In the same category of incendiary operations against shipping it might even be claimed that the Ancient Greeks were the pioneers with their introduction of the infamous Greek Fire. Attributed to the inventive brain of one Callinicus of Heliopolis, about the year 668BC, the ingredients of this highly inflammable substance have been lost in antiquity; but it must certainly have been a liquid and is now thought to have had a naphtha base with possibly phosphorus added, as it was squirted through tubes in the bows of boats propelled by rowers on to the decks of enemy ships. When suddenly enveloped in flames the enemy had no protection against such a fiendish device: to them this mysterious fire must have been like a modern attack with napalm bombs.

Although directed from a small boat to destroy a larger vessel, Greek Fire was air borne and not quite within the definition of mine warfare. Even the attack on French ships by the gallant Duke of Buckingham in the harbour at La Rochelle in 1628, using gunpowder-filled boats with gunlock igniters – on this occasion with more noise than damage to the enemy – could not strictly come within this heading. Unlike the lurking mine, these contrivances were more in the category of floating deterrents than of either torpedoes or mines.

It was, in fact, an American citizen, David Bushnell, who could be described as the father of the underwater mine or torpedo. He had discovered that, provided it was kept dry, gunpowder could be exploded under water. During the War of Independence in 1776 his native patriotism and inventiveness, added to the exasperating sight of English warships occupying his country's harbours, encouraged him to devise a way to place an explosive charge against the bottom of a ship, the soft underbelly where every ship is most vulnerable.

Bushnell produced a one-man submarine, not the first of its kind in history but the first to be constructed solely with the intention of sinking an enemy ship. His vessel, *The Turtle*, was roughly egg-shaped and was fitted with a hand-operated propeller and rudder to give it steerage way, a vertical propeller to enable it to hold its depth with some finesse, and water ballast which could be taken in or expelled as required by a hand pump. The boat was weighted also with about 600lb of lead ballast, and Bushnell arranged that one-third of this could be dropped so as to bring his boat rapidly to the surface in emergency.

The offensive part, the explosive charge, was a watertight keg containing some 150lb of gunpowder which could be released from inside the boat. Attached to the keg by a stout lanyard was a handscrew with which it was intended to attach the mine to the bottom of the ship. The operation of releasing the keg from the submarine boat started a clock running, designed to give the operator time to propel his boat clear before the gunlock mechanism was actuated.

As the whole operation was to be carried out by one man working on his own a commendable amount of Yankee skill and daredevil courage was called for, and this was not lacking. Amongst those who volunteered for the task a sergeant, Ezra Lee, was chosen, and an attempt was made on one of His Majesty's blockading ships then lying in New York Harbour.

Under cover of darkness Lee was towed to a point uptide of the British and released, but despite his efforts at the propeller handles he found the tide was carrying his egg past one tempting ship after another. Happily for his purpose he managed to close up with Lord Howe's flagship, the 64-gun line-of-battle-ship *Eagle*, and rose quietly against her bottom.

Twice he worked at inserting the handscrew so as to attach his charge, but the screw just could not get a hold, for he discovered that the flagship's bottom was completely copper sheathed up to the waterline. Dawn was not far off by now, and Lee realized that he would soon be discovered. There was nothing for it but to withdraw as silently as he had come, slipping the powder charge so as to make an easier getaway.

Lee's attempt was only the first of several others made against the blockading British. But the difficulty of holding the egg against a ship in a tideway while the operator tried to insert the augur into the vessel's tough bottom planking was to prove insuperable, and not one of these forays was successful. Whilst this method of attacking an enemy ship could be described as a forerunner of the limpet mine, iron and steel ships and magnet-attached explosives were as yet far away in the future.

Despite these failures, Bushnell in the following year, 1777, decided to try using a large keg of gunpowder supported by logs and fitted with an ingenious trigger-operated device. On this occasion he selected HMS *Cerberus*, a 32-gun frigate, as she lay at anchor just inside the mouth of the Connecticut river. Unfortunately for his hopes the strong tide carried his machine past the English warship, but it brushed its firing gear against the side of a small schooner anchored astern, and sank her within minutes.

Expectation still ran high for the effect Bushnell's explosive contrivance could have on the enemy, and the following January a large number of powder-filled kegs with trigger igniters were assembled on the banks of the Delaware river above Bordenstown.

Suspended about 3ft beneath wooden floats the kegs were set adrift on the tide so as to be carried down on to the long line of warships which stretched the whole length of the city.

Conditions, however, were not on the side of the Americans, for there was so much ice on the river that the British had wisely warped their ships into the wharves the previous evening, leaving the river comparatively clear; the ice also held back the run of the tide, so that daylight came before the first of the kegs reached the ships. Although the kegs themselves could not be seen their floats were soon spotted, and at the order 'Boats away!' the Jack Tars were soon engaged in destroying or capturing the kegs and rendering them harmless. Only one boat's crew who treated the Yankee invention with carelessness blew itself up, but for years afterwards the favourite yarns on the Lower Deck and in waterside taverns told of the Battle of the Kegs.

Robert Fulton (1765–1815), also an American citizen and a near contemporary of Bushnell, was better known as the originator of commercial steam navigation on the rivers of America with his steamboat *Clermont* on the Hudson River in 1807. He was, however, an inventive genius and in addition to his wide interests in steam engines and high pressure boilers, he put forward various proposals for underwater attack on enemy shipping.

During the time of the French Revolution the Napoleonic ports were being blockaded by the British fleet, and this gave Fulton an opportunity to try his inventions in practice. Improving in details on some of Bushnell's ideas in 1797 he built a submersible boat named *Nautilus*, with which he planned to carry an external explosive charge – he called it a floating carcase – to be attached to the bottom of an enemy ship, and offered the scheme to the French Government. But with a disarming display of altruism the Ministry of Marine turned it down on the grounds that striking an unsuspecting enemy from underneath was unethical and against all the rules of chivalry.

Undeterred by Gallic evasiveness Fulton offered his ship-sinking device to the Dutch Government, but was again rejected, whereupon he returned in 1800 to France and managed to enlist Napoleon's support together with a grant of 12,000 francs to build a second *Nautilus*. With this boat the following year he sank a small vessel which had been put at his disposal for demonstration purposes. But still the Ministry of Marine kept to their lofty disdain of such immoral practices, and with little outward sign of national bias Fulton crossed to England in 1804 and began lobbying the British Government.

Possibly not blinded with the same sense of chivalry the French authorities had displayed, Mr Pitt, the Prime Minister, showed great interest and appointed a special commission to offer Fulton facilities in demonstrating his claims. But although many experiments were carried out both with submerged explosive charges and with barrels of gunpowder on rafts floated down on the tide towards French warships lying off Boulogne, the commission remained unconvinced of the usefulness of this form of warfare.

As the forthright Earl St Vincent commented at the time, 'Pitt is the greatest fool that ever existed to encourage a mode of warfare which those who command the seas do not want, and which if successful will deprive them of it', and his words reflected the feelings of the Admiralty and probably of all the officers of the Fleet together. Shrugging the British off as blockheads after a few more attempts to win them over to his schemes, Fulton returned to his native America.

Here his activities were once more encouraged and a new committee was set up in 1810 to consider them. Fulton's inventions ranged from a gunpowder-filled metal torpedo slung beneath a float carrying stout wire outriggers which on brushing against a ship's sides would spring a gunlock firing mechanism, to what was to become the first true moored contact mine.

This comprised a cylindrical copper canister with hemispherical ends containing a 110lb charge of gunpowder. On the top of the charge case was soldered a small brass box fitted with a firing arm, which on being struck sprang a gunlock which fired into the main explosive charge. To give the canister the necessary buoyancy, or lift in the water, a wooden box filled with cork was attached to its sides like a saddle. The mooring consisted of a heavy weight with the required length of rope to keep the mine at the right depth so as not to become visible at low water.

The device incorporated in addition another of Fulton's ingenious inventions, by which the firing arm was automatically locked and rendered inoperative after the mine had been under water for any predetermined length of time from one day to a month or more. This mine assembly of 1810 therefore possessed many of the essential features of the moored contact mines in use ever since.

With the final submission of Bonaparte and the cessation of hostilities in Europe in 1815, and the death of Fulton in the same year, there was no longer any pressure on inventors to produce noxious methods of underwater warfare. Nevertheless a new name came on the scene to bring a new aspect to mine warfare.

For some years Colonel Colt, associated with the development of the revolver, had been experimenting with a method of firing moored mines electrically from an observation post on the shore. The problem was for the observer to find the exact moment when to detonate the mine. Colt found the solution by employing an electrical circuit which was closed if a ship touched the mine and transmitted the signal to the watching post. By this method he demonstrated that he could sink a moving ship more than four miles off the land.

By means of Colt's invention a system of *selective* mining became practicable by which important harbours could be protected by minefields from enemy ships seen to be approaching, while friendly or neutral vessels could be allowed to pass over the mines unharmed.

During the Schleswig-Holstein conflict with Denmark in 1848–51 the Prussians successfully used electrically controlled mines to protect Kiel harbour from the Danish fleet; but it was the Russians who first employed extensive minefields *strategically* when they laid minefields in defence of their Baltic harbours, Sveaborg, Kronstadt and Bomarsund, as well as at Sevastopol, during the Crimean War, 1854–56. Although they laid many mines on the sea bed with appropriately heavy charges to be fired electrically from observation posts ashore, the most numerous mines used by the Russians were of the moored contact type.

The design of these mines supplied to the Russians was attributed largely to the father of Alfred Nobel, of explosives fame, and incorporated the earliest known form of hollow lead horn attached to the outside of the mine shell. This horn enclosed a glass phial filled with sulphuric acid and embedded in a mixture of sugar and potassium chlorate. When a ship struck and bent the horn the glass phial was broken and the acid mixing with the chemicals produced a flame that ignited the main 25lb gunpowder charge.

Nobel's invention of the lead horn and acid-filled glass tube was later to be developed into the type of horned mine which served the governments of the world over more than a century of mine warfare. The mine had in fact arrived at its two principal forms of development – the moored contact type, and the remotely controlled electric mine – but it was to be during the American Civil War from 1861 to 1865 that mines were to play a major part with new inventions destined to make wars more devastating and horrible than ever before.

Chapter two

WAR OF INVENTIONS

With the secession of the Southern States from the Federal Union in 1861 America found a civil conflict on her hands which was to last over a full four years of bloodshed and misery. Compared with the almost unlimited reserves of the North, the secessionists possessed pitifully few resources in the form of iron foundries, ammunition factories, railroad materials, ship repairing yards, manpower, and everything needed to wage a long war. But what they lacked in the tools of war the so-called rebels made up for in determination, courage and ingenuity.

In the early stages of the War between the States the dashing patriots of the South won one encouraging victory after another against Union troops in the field, but gradually they became aware of the stranglehold that the Northern blockade was maintaining on all the Southern ports and rivers, depriving them of vital military stores and help from overseas. The Southern exploits on the battlefield and their splendid cavalry charges did nothing to break the ring of Federal gunboats that patrolled their coastline.

Having so few warships compared with the Yankee fleets, the Confederates were forced to find other ways of destroying, or at least disabling, the blockading enemy ships. In this they showed a high degree of inventiveness and personal courage, and proceeded to make extensive use of underwater mines, or torpedoes as these weapons were still called (the *locomotive* torpedo had not as yet been introduced).

At first these mines were simple kegs, well caulked so as to be watertight, filled with a comparatively small charge of about 22lb of gunpowder, and actuated by means of a simple friction igniter much on the principle of the Christmas cracker. The kegs were moored by a rope and a heavy ground weight so as to float just below the surface at low water, while their igniters were connected loosely in pairs of mines by a wire. On the trip wire being jerked by a passing ship both mines were exploded, and a number of Federal ships were damaged, and a few sunk, in this way. But the Northerners discovered that they could pass over the pairs of mines with a shallow draught vessel near high tide, and clear the channel by towing a ground chain with grapnels between two boats without undue difficulty.

For minefields at Charleston, Savannah and Mobile, and in the lower reaches of the Mississippi, various types of moored mine were laid in two or more lines athwart the channels. Most of these mines were again wooden kegs, some wired together with friction fuses, others with horns and Nobel-type chemical fuses which have already been described. An ingenious type which the Federals found very difficult to sweep because of its shape was one invented by a Lieutenant Brooks of the Confederate Navy. This consisted of an iron canister shaped like an inverted cone, on the dished base of which were five Nobel chemical horns, while the pointed end was firmly attached to a wooden spar. The other end of the spar was shackled loosely to a ring enclosing a cross-bar in the centre of a heavy ground weight. This permitted the spar with its mine to sway gently to and fro in the tidal stream with the base of the canister 3ft or so beneath the surface. Any passing sweep wire or chain would merely pass up the spar, pushing the mine away, and slide harmlessly over it.

Another variation was a large cylindrical iron canister holding in its lower half a 165lb charge of powder, and moored by a central iron rod and lifting eye and rope to a suitable anchor or weight. The top of the canister held a heavy loose cover which was attached by a length of loose chain to a plug and friction fuse in the base of the charge. Upon the mine being struck by a ship the cover was dislodged and, falling, jerked out the plug and actuated the fuse.

To prevent the mine from being fired accidentally while being laid, and also to assist recovery when the mine was no longer required, there was a second, but shorter, length of chain both to the cover and to a safety pin which was inserted through the central rod below the canister. This was designed to take the shock of the falling weight instead of the firing mechanism. Once the mine was laid the safety pin was withdrawn by a gentle pull on its lanyard, and the mine at once became active. For recovery the pin had to be reinserted in its hole so as to render the mine safe. To appreciate how well thought out this mine was it is true to say that the drill for laying, arming and recovering fulfilled almost all the safety regulations laid down for handling moored contact mines to this day.

In the latter half of the war the Confederates developed various types of ground mines, firing them electrically from observation posts on shore. Any explosive charge to be fired at some distance from a ship has to be a large one if it is to sink or seriously damage the vessel. The Southerners were fully alive to this and in more than one instance at Charleston and Savannah a condemned locomotive boiler with firebox and tubes removed was used as a ground mine and filled with as much as 1300lb of gunpowder.

The kind of subterranean upheaval such a charge would create in a few fathoms of water might be spectacular, but unless it was within some 60ft of a strongly built ship it would only shake her, not necessarily sink her. As the Russians had found during the Crimean campaign, with all controlled mines fired from on shore the exact moment to fire when an enemy vessel was directly over the mine was very difficult to gauge. The Confederates were well aware of this and used various systems, from cross bearing with each observer operating a firing key to the electric contact method invented by Colt. The latter had the advantage that it could be used at night or during thick weather, but against that the ship had actually to touch a mine before the electrical circuit was completed. There was no easy solution as yet to this problem.

Although the Southerners were able to claim a number of outstanding successes with their minefields in sinking Federal warships, one of the recurrent problems, as with all electrical devices used under water, was that of keeping all electric cables and contacts in order. On more than one occasion when an enemy ship was observed approaching a series of shore-controlled mines, and the moment came to explode them, it was exasperating for the watchers when their frantic efforts to make contact and blow the Yankee ship out of the water passed as a silent non-event, and they could only stand helpless and watch the enemy steam steadily on her way. As one Yankee sailor was said to have commented, 'Gee, there sure was a lot of red faces amongst them Johnny Rebs!'

The unreliability of these early mines, electric or with striker fuses or chemical horns, could reduce to some extent the risk of steaming over them the longer they had been in the water. Off the harbour of Mobile, Alabama, three lines of some thirty mines each had been laid, incorporating a mixture of both the Nobel horn type and the drop weight design described above, and had been ready for the past three months for the expected attack on the harbour.

When it came it was led by Admiral David Farragut in his flagship *Hertford*, his attack force being preceded by four ironclad monitors which were considered strong enough to pass over any of the rebels' exploding mines without suffering unduly. In the event, the monitor *Tecumseh* in the van set off a mine under her bottom and sank in a few minutes with most of her crew. The second monitor *Brooklyn* at once reported that she had sighted more mines and signalled a warning to the flagship. But Farragut was made of stern stuff and, so the story has it, gave an order which has resounded in the histories of naval engagements ever since: 'Damn the torpedoes! Captain Drayton, go ahead. Mister Jowell, full speed.'

The Admiral's confidence was justified, for his ships forged ahead together, and although some of his captains later reported that they actually felt 'torpedoes' scrape along their ships' sides, none of them exploded. To the chagrin of those watching from on shore corrosion had seized up the firing mechanisms; the mines had been too long under water, and as a result Mobile fell into Union hands.

In faraway London the early admiration for the Southerners who were seen to be fighting for their principles against a more powerful force, was becoming a little eroded by increasing criticism of their use of so many mines. In the eyes of a nation proud of its sea heritage and of the British Navy, the scattering of lurking mines in huge numbers appeared to go beyond the rules of chivalry, and was far from being sportsmanlike. Britain had not then been at war on a substantial scale since Napoleonic times, and underwater warfare had not yet assumed any significance beyond the few losses suffered in the Crimea from mines laid by those unprincipled Russians.

On the other hand there were those who held that there was nothing unethical in a smaller and weaker nation fighting a greater using any available weapon, plus the element of surprise (recalling David's sling against the giant Goliath), to gain a victory. From the episode of the Trojan horse to the first introduction of the gun in the field, this element of surprise was thought by many to justify the means in any war.

The two schools of thought would continue to differ through the decades to come, but the next development in underwater strategy produced by the Confederates earned the admiration, wholehearted or grudging according to the viewpoint, of the British people, for it was not only a method of going out and *attacking* the enemy, but it also called for a high degree of endeavour and bravery on the part of the operators. So desperate had become the plight of the Southern States through the seemingly impregnable blockade of their coastline that the Confederates were willing to accept almost any personal risk in attempts to break it and enable supplies for their war effort to be brought in.

The result was the construction of special small craft which could carry a heavy powder charge up to an enemy ship and explode it under the turn of her bilge. Soon popularly known as 'Davids', these puny craft were in fact an updated version of a previous American invention already noted, Bushnell's submarine egg. The Davids were not, however, submarines in the sense that they could submerge completely, but by means of taking in water ballast they could be made to run in smooth water with the deck awash and only a low open hatch showing above the surface.

The powder charge, initially some 130lb, was carried in a metal canister at the outer end of a wooden spar pivoted at the vessel's bow so that, on the final approach to the enemy ship, the charge could be lowered beneath the waterline and exploded against the ship's bilge by a trip wire and friction fuse or a percussion mechanism. The first Davids were small cigar-shaped vessels propelled by the crew who turned the handles of a two-bladed screw. The first surprise attacks in 1863 achieved a measure of success in that the Federal flagship *Ironsides* was damaged at Charleston, the *Housatonic* sunk soon after, and the *Memphis* and *Minnesota* also so damaged as to be in dock for many weeks.

The method of attack, however, was hazardous in the extreme for, if the creeping submersible was spotted on its approach in the dark towards the Federal target, well-aimed gunfire could soon knock it out. On the first operations, also, the inrush of water caused by the explosion drew the David's bow into the hole, and the attacker went down with the ship. Yet despite the near-certainty of being drowned in the attack, after the boats were salvaged there was no dearth of volunteers for this hazardous service, and it was this cheerful bravery that captured the imagination of the watching nations.

The very presence of these Davids with their unpleasant habit of approaching unheard and unseen kept the Northern crews in a jittery state, and induced the captains of the ships in harbour to arrange all kinds of protection around them in the form of floating logs, spars lashed together and nets – the last a safety precaution which was to be repeated in later wars with the introduction of the locomotive torpedo.

With deliberate obstructions of this kind to break through the Confederates built larger Davids propelled by steam with a much greater radius and more speed. When running awash with the ballast tanks full only the stumpy smokestack, the engine room hatch and the enclosed pilot's lookout could be seen above the water. Presenting such a low profile these craft, like their hand-propelled predecessors, had little reserve buoyancy to keep them afloat should even a small quantity of water be shipped down the hatch; but it was essential to keep the hatch open for the inflow of air required for the boiler fire and the sweating crew. These steam Davids were in effect the forerunners of the ill-fated 'K' class steam submarines which were to be introduced in the British Navy during the 1914–18 War with such disastrous results.

Although the new steam Davids were a little faster and had a wider range than the hand-driven boats could manage, they were still liable to be carried down by their sinking victim. To try to avoid this, crews became expert in deciding when to put the engine into full astern, which was usually a few seconds before the torpedo made contact and the boat was losing her forward way. If smartly carried out – and the engine did not stop on dead centre! – the boat backed off before the inrush of water began and managed to escape unless sunk by gunfire.

At best the dice were heavily loaded against the volunteers of the Davids' service, and it became evident that a greater chance of success could be achieved with small *surface* craft. A number of fast steam launches were accordingly built which were capable of making a much speedier approach towards the enemy. They were equipped with the same type of movable boom and torpedo as before, and their successes during the last year of the War roused the interest of other navies.

Mobile spar torpedoes became widely adopted and were used in various parts of the world until, by various stages, the self-propelled or locomotive torpedo was invented, bringing in its train the torpedo boat to launch it and the torpedo boat destroyer. And in due course history was to repeat itself some eight decades later during World War II with the introduction of self-destroying speedboats carrying heavy explosive charges, and midget submarines.

This might appear to have been something of a diversion from the subject of mines, but the distinction between what constitutes a mine, whether moored, contact or non-contact, or resting on the sea bed, and the explosive charge that is conveyed to an enemy ship, is debatable: the one type easily leads to the other. It could be said that with the War between the States the underwater mine had come into its own as an acceptable weapon of war, having accounted for the loss on both sides of 28 ships in all.

In other ways, too, this American conflict could be described as the first of the modern-style wars, in the sense that it introduced new technology using up-to-date inventions. Amongst these innovations were the field telegraph, electric lamp signalling, captive balloon observation posts, hydrophones to detect the approach of enemy ships to a harbour, controlled minefields, ironclad monitors, heavy mobile mortars carried on railroad cars, and the strategic use of railroads for transport of troops and stores. And, sadly, this war inaugurated concentration camps for prisoners. The use of mines, not only individually to sink an enemy ship, but also in large groups as a deterrent against the movement of enemy fleets, had been shown as a new and effective power in sea warfare. This fact had not been lost on the nations of the world, but two other inventions were about to bring great changes in the old techniques.

For some five centuries the only explosive available for use with guns, bombshells and petards had been gunpowder. Over the years there had indeed been a gradual improvement in its quality, and the coarse-grained black powder emitting great clouds of smoke with which the Elizabethan ships had fought the Spanish Armada had led to the fine-grained brown powder which gave more power with less smoke by the 1860s. But the essential ingredients – saltpetre, charcoal and sulphur – were little altered, and the rate of combustion was still comparatively slow.

With the invention in 1866 by Alfred Nobel of an explosive compound to which he gave the general name *dynamite*, a new era of detonation could be said to have arrived; gunpowder was to give way to explosives of three, four and five times its power of destruction. Almost contemporary with this breakthrough in explosives the second innovation was to cause the navies of the world to rethink their future strategy in underwater warfare: the invention by Robert Whitehead of the self-propelled torpedo.

For the first time it had become possible to aim an unmanned explosive charge and for it to travel under water towards a *moving* enemy vessel. There had already, of course, been various suggestions for doing this, but hitherto without success. One version, for example, invented by Captain Harvey of the Royal Navy, consisted of a hollow float holding an explosive charge which was towed into position by a steam pinnace. On approach to a hostile warship the float or torpedo could be veered against the side of the enemy and the charge exploded by a trip wire from the pinnace. Although tried in a variety of forms, Harvey's towing torpedo was never adopted, for it had the great disadvantage that the vessel towing it was compelled to come close enough to the enemy's guns to be sunk before the torpedo could do its damage.

The torpedo devised by Whitehead, however, ran under water on its own power and, in theory at least, maintained a preset course and depth below the surface, and appeared to be the perfect weapon of surprise. The locomotive torpedo, with it warhead packed with 220lb of dynamite or guncotton, was like an uncanny, inspired mine which would propel itself straight through the water to crash itself against an enemy and blow a huge hole in the hull.

How the Whitehead torpedo was at first to encounter some fierce opposition to its adoption by the Royal Navy on the grounds that it could make nearly all the ships of the Fleet obsolete overnight ('and a demmed cowardly way of waging war, by gad!' growled at least one sea-roughened voice); how it was quickly seized as the ideal weapon by lesser powers with small navies; and how, like all horrible innovations for war, the torpedo was eventually accepted by every navy in the world as being as necessary as the big gun, need not concern us here. The torpedo rapidly grew out of its early stages as a hesitant, often irresponsible and highly unreliable self-propelled mine, and became a power of its own in the naval wars to come.

Chapter three

THE MINE SHOWS ITS TEETH

Contrary to popular belief, explosives cannot destroy completely; they may disrupt, shatter, burn or even disintegrate materials, but the original matter remains. Experts can usually piece together the fragments and in a short time give details of the explosive used, its quantity and type of fuse employed. Evidence is exceedingly difficult to erase merely by blowing it up.

An explosion is essentially an intensely rapid expansion of hot gases, which do their destructive work partly by the heat they generate and partly by the speed with which they strike their object. Explode a charge in the open and it does little more than make a loud noise: confine it within solid objects and the damage done will be related to the pressure built up by the expanding gases.

Gunpowder can be ignited by the simplest means such as a naked flame, a slow fuse, heat from chemicals or sparks from a friction fuse. Gunpowder has therefore always been a dangerous mixture to have to stow in quantity, especially aboard ships, and it must be treated with care. A high explosive such as dynamite, on the other hand, cannot normally be blown up by these simple means, but requires to be given a very sharp shock: in short, it has to be *detonated* rather than set fire to. In fact, like many of its variations dynamite can be burnt in small quantities (it makes excellent fire lighters), but if too great a mass is burnt at once the centre can reach a critically high temperature and explode. Burning explosives should never occur in excessive quantities nor in confined spaces, or the operator would be well advised to remember an urgent appointment somewhere else!

There are numerous varieties of so-called high explosives which rely on chemical reaction for activation, and their gas expansion rates vary from comparatively slow to very high. Those with the slower rates of detonation – in the 6000–12,000fs range – are suitable for use as propellants, as in guns, rockets and charges needed to shift earth and rocks. An explosive with a much higher rate of detonation would deliver too sharp a crack which could burst a gun barrel, while in breaking up rocks and moving tons of earth its too-violent expansion would not have anything like the same effect.

Amongst the high velocity explosives, examples are trinitrotoluene (TNT), trinitrophenol (picric acid), and pentaery-tritoltetranitrate (TNT/PETN) whose detonation rate is of the order of 18,000–27,000fs. To activate all these types of high explosive a separate small charge is required to give the main charge the sharp crack it needs to detonate it. This detonator usually takes the form of a small tube containing a composition which is sensitive to percussion or to heat supplied by a burning fuse or an electric spark. The explosive commonly found in detonators and percussion caps is mercury fulminate, a crystalline salt which explodes violently when subjected to friction or heat.

Remembering the number of warships which have been lost during the centuries through fire reaching their powder magazines, it is essential for explosives stored in any quantity, aboard a ship or on land, that they should be as safe as possible. This calls for an *inert* type of explosive, manufactured so that a spark, a modest fire, or even a collision at sea with its attendant sparks, should not cause it to blow up.

By mixing the necessary chemicals with a suitable base, such as charcoal or kieselguhr, a high explosive is rendered sufficiently insensitive for it to be safe to handle and to be transported in quantity. To initiate a heavy charge of one of these inert explosives a small detonator is not enough. It would often do no more than blow a hole in the main charge, which would fail to detonate. During World War II the Germans used a particularly inert explosive called hexanite in their larger mines and bombs, and many cases of only partial detonation, sometimes complete failure to explode, were reported.

To fire a heavy main charge a two-stage method of initiation is necessary, and this is achieved by using an intermediate charge, called the *primer*, into which the detonator is inserted. This priming charge is usually a cylindrical metal case containing an explosive not so insensitive as the main charge but still reasonably inert. The mercury fulminate detonator would initiate the primer, which in its turn would give the main charge the shock needed to explode it.

It is safe, therefore, to store and handle mines, torpedoes and bombs with large charges provided they have not yet been primed. When the primer is inserted the object may still be moved with reasonable care, but without fear of a premature explosion; but when the detonator with its electric leads or length of fuse is inserted in the primer the object becomes *armed* and is ready for firing when required.

By constant experiment the design of the mine had by now reached a form which would not be improved upon for several decades. With the adoption of more powerful explosive charges the old type of horn, attributed to Nobel – which as we have seen mixed chemicals on being bent, to produce a flame and ignite the gunpowder charge – was no longer suitable. Its place was taken by a new type, invented by Dr Otto Herz in 1868, when he was working with the German Naval Defence commission.

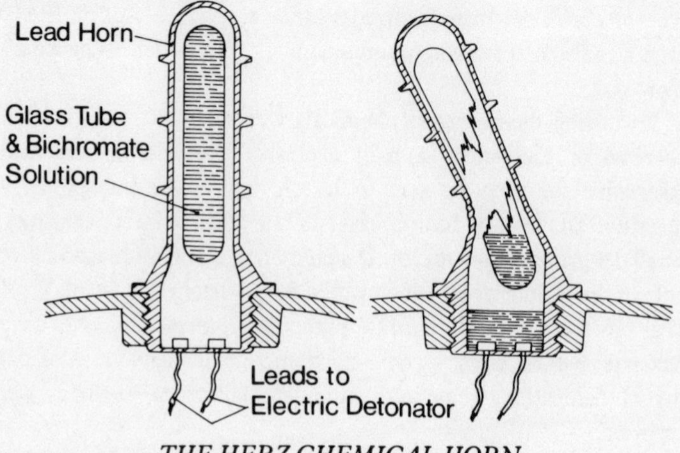

THE HERZ CHEMICAL HORN

In the Herz horn a bichromate solution in a glass tube dropped on to carbon and zinc plates in the base of the horn unit when the glass was fractured. This mixture formed a battery cell which produced sufficient current at about 1.5 volts to fuse a fine platinum wire embedded in the mercury fulminate of the detonator, which in its turn initiated a guncotton primer and so fired the main charge.

The beauty of the Herz horn was that it was virtually foolproof, and it incorporated all the necessary components of a dormant battery, which would not spring to life until the glass tube inside the lead horn was fractured. Mines fitted with this type of horn therefore had an almost indefinite life, for there were no batteries to need recharging. The Herz horn was to remain the most widely accepted method of exploding contact mines for the next seventy years or more.

With improvements in contact mine design came better methods of laying the mines. Hitherto the problem of fixing the length of mooring rope so that at low water the mine would not be so close to the surface as to be spotted was met by calculating the state of the tide when the mines were to be laid, and adjusting the lengths of all the mooring ropes accordingly. This required a careful survey of the sea bed and a knowledge of the depths of water at all states of the tide; it could indeed be a tedious operation if the laying had been delayed for any reason and the lengths of the mooring ropes had to be readjusted to meet the changed tidal conditions. If the minelayer's calculations were inaccurate some of the mines could be left 'watching' at low water, that is breaking surface and visible to the enemy, who could then easily deal with them.

The problem was an acute one, and after HMS *Vernon* had been commissioned as a separate torpedo establishment at Portsmouth in 1876, experiments were concentrated on these factors. An ingenious solution, attributed to Lt Ottley, a *Vernon* officer, was the automatic or plummet sinker. This was designed to enable mines to be laid so as to be at a calculated depth below the surface whatever the state of the tide at the time of laying.

With this plummet system a separate weight is allowed to run out on its wire to a predetermined length, which will be equal to the depth that the mine is to be moored beneath the surface. As the mine and its box-like sinker are dropped over the stern rails of the minelayer, the sinker descends towards the sea bed with the plummet weight on its wire hanging below it, while the mine remains on the surface. The tension on the plummet wire holds a spring pawl away from the mooring wire drum, which is free to revolve inside the sinker. The moment the plummet reaches the bottom the tension on its line is released, the pawl engages, and the drum is stopped from revolving. The sinker settles on the sea bed, drawing the mine down to its proper depth.

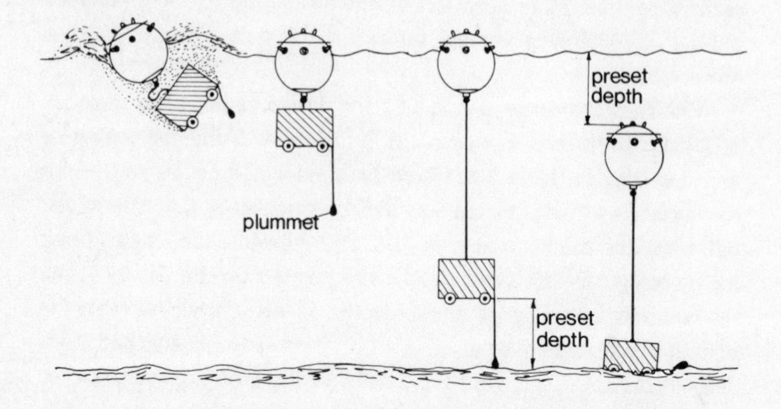

AUTOMATIC DEPTH-SETTING WITH PLUMMET

Like all innovations for use at sea the plummet system had many teething troubles in its early stages. It was found, for instance, that as the sinkers fell away from their mines before hitting the surface of the water some managed to unwind more mooring wire from their drums than others, and the mines accordingly would take up varying depths below the surface.

This was especially so if minelaying took place at speed; but speed of laying a minefield near to an enemy coast is just what the minelayers want most of all. Despite its drawbacks, however, the plummet system became widely adopted by most of the world's navies, until a more accurate method of depth-setting was to be invented many years later, using a hydrostat device actuated by the pressure of the water.

During the 1890s the Admiralty was reluctant to follow any active policy on mining. It was presumably still considered in high places hardly necessary for the world's principal naval power to stoop to such underhanded methods as luring the enemy into minefields. For the same reason, it could be said, the Navy was at first slow to follow the other Powers in the laying down of submarines. The War Office, on the other hand, was insistent on augmenting their own defences of the principal ports around Britain and the Empire with groups of controlled minefields, and the naval authorities accordingly obliged with the production of suitable sea mines. These were generally of two types: observation mines moored in lines and fired from a battery on shore, and a smaller moored electro-contact type of mine laid in groups in the harbour approaches.

When war between Russia and Japan broke out in 1904 the mine menace raised its horned head with a renewed vigour. Both sides laid large quantities of contact mines with plummet sinkers, and the Japanese strategic use of extensive minefields in influencing the movements of Russian ships demonstrated what a decisive factor this could be.

The Russian mines were mostly of Herz horn type with guncotton main charges and their own design of safe-handling device. This employed an electric switch on a cable on the outside of the mine shell, which showed all the regrettable traits of any mechanism immersed for long in seawater: it was responsible for many premature detonations, some while being recovered.

The Japanese mines, as might be expected, were on the whole more ingenious. Spherical in form without any external horns, the mine contained a 110lb charge actuated by a heavy inertia weight which, on the mine being struck, swung against an all-ways contact plate wired to the detonator. While being laid the mine was held safe by a clockwork timing mechanism in conjunction with a soluble plug, giving it a delay of about half an hour before the mine came 'alive'. What arrangements were made for rendering these mines safe while being recovered, except by sinking or countermining them, is not known.

The comparative ease with which the Japanese laid minefields off the Russians' principal ports of Dalny and Port Arthur, and then cunningly lured units of the Russian fleet into a position where the latter would have to retreat through the mines on the unexpected appearance over the horizon of the main part of the Japanese fleet, was not lost upon other nations. This manoeuvre took place in May 1904 and resulted in the sinking of the 11,000-ton flagship *Petropavlovsk*, whilst the 13,000-ton *Pobieda* was so damaged as to be out of action for some months.

The Japanese did not, however, have it all their own way, for they shortly afterwards suffered the loss of two of their principal battleships, *Yashima* and *Hatsuse*, through a skilful operation of Russian minelaying carried out the night before. Altogether in the Russo-Japanese War the two opponents lost between them three battleships, six cruisers, four destroyers and six smaller units through mines alone. It was a significant enough indication that mining had become a highly effective weapon in sea warfare, and for the other nations of the world the writing was already on the wall.

The British Admiralty was no longer slow to take notice, and the development of a standard mine for home and Empire protection was put in hand at once. The result was designed to supersede the old harbour approaches observation and electro-contact types, and in 1905 an order was placed for one thousand mines of the new type.

Known as the British Spherical, this was a moored contact mine using the latest plummet sinker. In place of Herz horns the 110lb guncotton charge was actuated by a horizontal firing arm pivoted on the top of the mine case. On this arm being turned a spring-loaded pin struck a detonator which fired the primer and the main charge.

At first the firing mechanism was prevented from operating until after the mine and sinker had been laid and had come to rest in position, by means of a soluble plug which contained ordinary sugar.

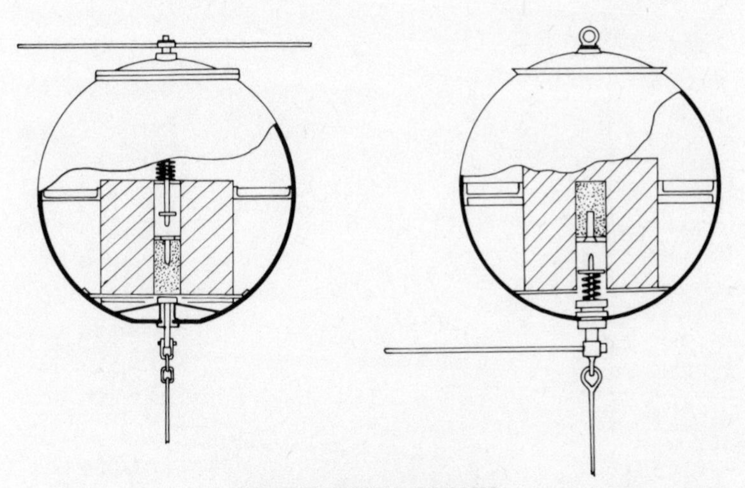

EARLY NAVAL MINES
The British Spherical, 1905 (left), and the British Elia, 1914 (right)

Some consternation was caused when on one occasion a mine exploded almost as it struck the water, fortunately with only minor damage to the minelayer and no casualties amongst her crew. Enquiry suggested that ratings with a sweet tooth had removed the sugar filling, so that the firing arm could be twisted round on striking the water. Replacement of all soluble plug fillings with a less tempting substance – in this case salammoniac – cured the trouble and no further 'prematures' were reported.

By the time the First World War broke out in 1914 British minelaying policy had been given high priority. In addition to the prewar stocks of the standard naval spherical mines an improved version of an Italian design, to be known as the British Elia, was ordered to be delivered in large quantities. This mine was similar in operation to the earlier type, with a firing arm releasing a spring striker, but in this case the bar was pivoted at the base of the mine.

For use in shallow enemy waters where the tides tended to flow in only one direction, such as along the western coasts of Germany and Denmark, other ingenious forms of mine were made. These included three principal types:

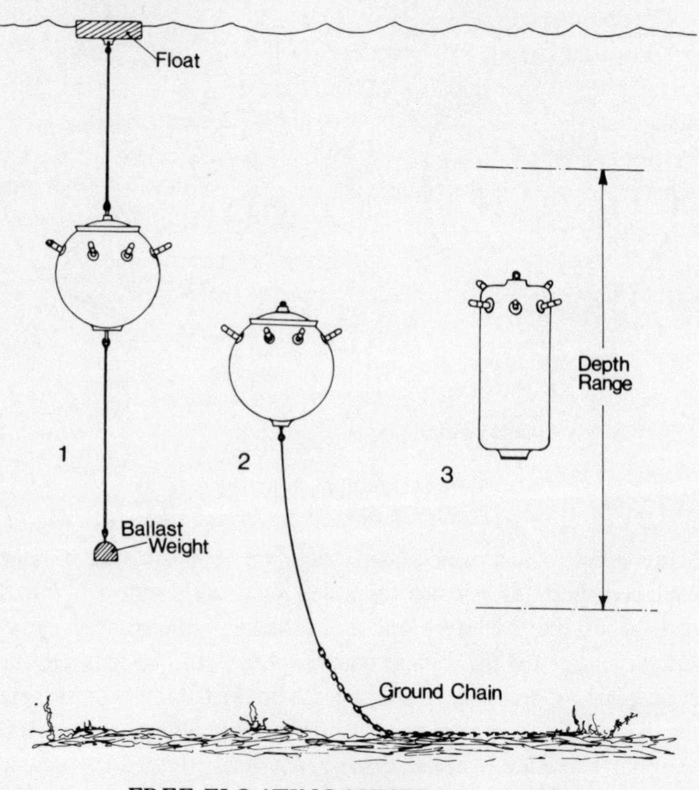

FREE-FLOATING MINES (not to scale)
1. Drifting
2. Creeping
3. Oscillating

(a) a *drifting* mine which hung from a flat and not easily spotted float on a short line, having a ballast weight to keep the mine from rising to the surface;
(b) a *creeping* mine attached to a length of chain which dragged along the bottom and held the mine at a roughly predetermined distance above the sea bed; and
(c) an *oscillating* mine which was free floating but designed to rise and fall slowly within some range of depths below the surface.

The operation of the last-named type was simple enough, as its up and down movement was controlled by a hydrostatic valve which worked a displacement piston inside a cylinder of ammonia gas. The mine would rise to within a few feet of the surface, the reduced water pressure would release the valve, a small quantity of the gas would escape, and the mine having lost this buoyancy would slowly descend, until the cycle repeated itself or the gas became exhausted.

Previous to this bobbing drifter there had been an even more ingenious type of oscillating mine produced in Sweden. Known as the Leon, this was a cylindrical mine having a guncotton charge of some 150lb which was fired electrically by an inertia mechanism. When dropped overboard, or released from a submarine, the mine sank slowly until at a predetermined depth a hydrostatic valve closed a circuit from a battery to a small electric motor which drove a propeller in the base of the mine. The mine then rose slowly towards the surface until the hydrostatic valve again broke the contact, the motor stopped, and the mine began to sink again to repeat the cycle.

The process would continue with the mine ranging up and down within its intended range of depths until switched off by a timing device or until the battery ran down. Effective though it undoubtedly was against submarines as well as the unfortunate surface ship which might strike the mine on its approach to the surface, the Leon was not adopted by the Admiralty; in fact little use was made of any of these schemes for drifting mines as it was apparent that such uncontrolled devices, once let loose, could be as much of an embarrassment to British ships and submarines as to the enemy's.

By 1914, the long-nurtured reluctance of the Admiralty to sow mines in the path of the enemy was being eroded by the turn of events in Europe and by the threat of German U-boats in the Straits of Dover. Vigorous plans were accordingly mounted to lay extensive minefields off all the principal ports of the east coast of England and Scotland where enemy raiders were expected to operate.

In accordance with International Law, notices to mariners were issued promptly, giving the positions of the danger areas, while neutral shipping as well as convoys were duly routed between the minefields and the coastline. Later on in the war the Admiralty was to proclaim the whole of the North Sea (the German Ocean of prewar school atlases) to be a danger war zone.

The need to manufacture the thousands of mines required put a severe strain on a country already producing guns and shells in huge quantities, whilst the shortage of ships that were suitable for laying all these mines was acute. While specially designed fast minelayers were being hurriedly built in the shipyards, a number of cross-channel and passenger/cargo steamers were taken over and converted for minelaying. These relatively fast ships were capable of taking some hundreds of mines each; they were fitted with parallel lines of rails on the main deck ending in a sudden dip over the stern, over which the wheeled sinkers ran, each one carrying its mine like the world on the shoulders of Atlas.

Gradually as mine production was stepped up with all its associated equipment – mine shells, charge cases, firing mechanisms, sinkers, launching gear, and thousands of miles of mooring cables – the newly equipped minelayers played their part in sowing minefields along the English Channel coast, in the Heligoland Bight, and off the enemy's own shores. But despite these precautions the Germans were still not prevented from sending their U-boats out into the North Sea and round the north of Scotland into the Atlantic, from the Weser and the Elbe and from out of the Baltic by way of the Skagerrak. Already their effect on home-coming convoys was growing serious.

It was not until America decided in April 1917 to commit her vast resources of men and materials to the war in Europe that a really effective policy of mine warfare became practicable. With a far greater production of mines and mining equipment becoming available with supplies from across the Atlantic, the minelayers were able at last to complete the closing of the Dover Straits to U-boats with deep- and shallow-laid mines as well as mined submarine nets. In addition, protective minefields were added at the approaches to the Irish Sea and at many other strategic points around the coasts of Great Britain.

If ever a lesson were needed on the value of mining waters where enemy ships were expected to operate, it was driven home to the Admiralty during the first week of June 1917. The Russians had been making pressing demands on Britain for both financial help and military supplies to be sent them to enable their army to stave off the German advance on the eastern front. (History was to repeat itself with a comparable situation a quarter of a century later during the Second World War, resulting in the problems of the Arctic Convoys.)

The Russians' expectations were embarrassing enough at the crucial stage the war had reached in 1917 for the Cabinet to decide to send a delegation headed by no less a person than the Secretary of State for War, Lord Kitchener, whose firm personality would be adequate to convince the Tsar of the difficulties the Allies themselves were facing in the west. The meeting was due to take place in St Petersburg, and the party of high-ranking VIPs joined the cruiser *Hampshire* at Scapa Flow on 5 June en route for the port of Archangel at the head of the White Sea.

The weather had been unseasonable and stormy for a week, and was so bad with a gale screaming through the naval anchorage from the north-east, that when the cruiser with her two destroyer escorts got under way that evening, instead of steaming south-east about Ronaldsay and on to her intended north-easterly course for northern Norway into the teeth of the wind, the ships were ordered to proceed west about. This course would give them a lee from the islands and enable the destroyer escort to keep up with the *Hampshire*.

No sooner had they brought the pillar rock, the Old Man of Hoy, abeam than the wind backed suddenly into the north-west and blew great guns right on the nose. It would have been better, as it now turned out, if the original plan had been adhered to. In these sea conditions the two destroyers could not keep up with the speed of the much bigger cruiser, and they were accordingly ordered back to Scapa Flow, while the *Hampshire* proceeded on her north-westerly course alone, keeping some 2–3 miles off the shore.

Great events had indeed been happening in the area during the last week in May, for the combined German and British fleets had met and engaged one another in what proved to be the greatest fleet action of the war, the Battle of Jutland – and it was during all this activity that a minelaying U-boat had rounded the northernmost of the Isles of Orkney and, undetected by any British naval vessels, passed down the coast of the main island until a selected point close to Marwick Head was reached. From here towards the west her cargo of mines was laid in five groups of four mines each.

The Germans had long observed that all British naval vessels passing west of Orkney were in the habit of hugging the Orcadian shore to within three or four miles. The U-boat's mines were therefore laid to cross this approximate fairway, and were set at a depth which would not affect any of the smaller fry – the trawlers, minesweepers or even destroyers – but would hopefully catch one of the bigger units with a draught of 25ft (8m) or more. Having laid his eggs the German returned to base.

The *Hampshire* had been less than four hours on her way to Archangel when almost abreast of Marwick Head she struck one of these mines. From the handful of men – only a dozen of them – who managed to make the shore on liferafts and survived it was learnt that there was only a single explosion, but a very heavy one. The cruiser almost at once rolled over to starboard, and, said one, 'My God, the old girl was gone in a few minutes!'

An ordinary contact mine striking a cruiser like the *Hampshire* at the bow would not normally have been enough to sink her; she would have been disabled and have had to return to base decidedly down by the head, but still operable. It seems likely that this mine was set deep enough for the ship to have passed partly over it and in the big sea running her bottom came down and touched it off, for the explosion was described as amidships. This caused either one or more of the boilers to blow up or one of the magazines, with fatal consequences.

Apart from the twelve men who managed to get ashore there were no survivors, and Great Britain suffered one of the most profound losses of the war in the death of the great soldier-statesman. No other delegation was appointed to make the journey to St Petersburg, and in little more than a year the Bolshevik Revolution swept over an unhappy Russia, and the Tsar was no more.

At first it was assumed that the *Hampshire* had been torpedoed by a U-boat cruising in the area, but from the elated German press and, later on, from official reports it was realized that this tragic disaster was the result of a mine. It may be idle, but at least it is thought-provoking, to speculate on how different might have been the history of the war had this mine not done its deadly work, and had Kitchener been able to return from his mission and continue to exert his influence in Whitehall.

Despite the effectiveness of the Folkestone to Cap Gris Nez mine barrage, which included in its length sections of both mined anti-submarine curtain nets and controlled loops, U-boats were known to be still slipping through from their bases in the Elbe and the Weser. This route down Channel was so much shorter than the long haul round the North of Scotland to reach the Allied convoys in the Atlantic that it was decided for the U-boat commanders that the risks of finding a way through the barrage were justified. And in consequence the enemy's harassment of the Atlantic convoys in the approaches to Great Britain was becoming more and more grave.

A determined scheme was accordingly put in hand to seal up the Dover Strait completely against both U-boats and surface raiders. This astonishing plan consisted of nothing less than the planting of a row of massive concrete towers at 3-mile intervals across the Channel. Each tower, like a small fortress, was to be equipped with searchlights and guns and to act as a control centre for the minefield loops laid between each pair of towers.

Had this plan been completed it would undoubtedly have formed the most effective barrage against enemy marauding, (even though friendly ships' captains might have expressed strong opinions of the hazards to navigation such towers would have presented in thick weather, much as they were to protest many years later at the proposal of a Channel Bridge) but the war came to an end when only the first of these towers had been completed. Towed out from its building berth at Shoreham, it was finally sunk in a position some five miles east of Bembridge in the Isle of Wight, and as the Nab Tower it has been a prominent seamark for the entrance to Spithead ever since.

In the meantime the fleets of U-boats which the German shipyards were building in ever-increasing numbers were reaching the Atlantic with little hindrance, despite the greater distance of their northabout route. The consequent losses of merchant ships with their desperately needed cargoes were becoming so great in the second half of 1917 that those in the know realized that if no effective counter-measures could be put in force, Great Britain was in real danger of defeat through strangulation. Starvation of her people and deprivation of essential war supplies would sound her death knell.

Plans to try to bottle up the enemy submarine fleets in the North Sea, to prevent them from emerging from their bases in the Baltic and reaching the Atlantic, were put in hand. Anything as closely knit and effective as the Channel barrage would not be practicable because of the distances involved, but it was decided to close off the northern section of the North Sea by a huge barrage of mines which would be laid all the way from the east coast of Scotland to the coast of Norway.

The magnitude of the plan was such that the numbers of mines that would be needed, together with their sinkers and mooring wires and all the related equipment, would be quite beyond the resources of British manufacturers. Nor were there sufficient minelayers of a suitable type available.

It was here, in the eventual laying of this great northern mine barrage, that the tremendous production potential of the United States showed itself, and made it possible. The final plan decided upon was for lines of both deep and shallow-laid mines to form a barrier from Scapa Flow in the Orkneys to the edge of the Norwegian minefields north of Egersund, a distance of almost 250 miles.

From the strategic angle the siting of the barrage finally agreed was a brilliant conception, for it would not only close off the entrance to the Baltic and all the enemy's ports, but it would be too far for German minesweepers to operate without adequate support, and it was thought that would hopefully tempt the German High Seas Fleet to come out and find themselves again in action with heavy units of the Grand Fleet. Furthermore, based at Rosyth and at Scapa Flow, the Navy would be on the enemy's side of the barrage, where patrolling of both ends would be easier while Norway remained neutral. As a final bonus for the scheme any U-boats mined, but only disabled, in the barrage would have a slim chance of making a home port.

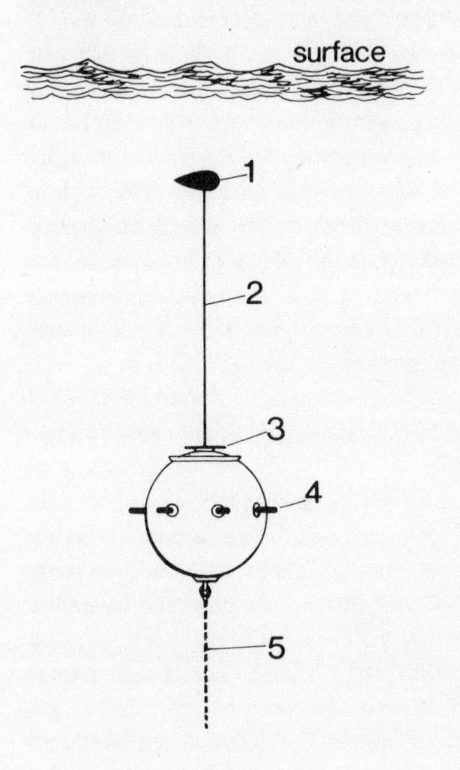

1. *Float*
2. *Antenna*
3. *Copper plate*
4. *Switch horns*
5. *Mooring wire*

ANTENNA MINE (not to scale)

At first it had been planned to lay mines at varying depths so as to catch U-boats running on the surface as well as submerged, but the numbers of mines that would be required to complete such a length of multi-layer field was astronomical, and for this barrage a new type of contact mine was devised. This was the *antenna* mine, designed essentially to catch a passing submarine, or a surface vessel, over a considerable range of depth.

In the form adopted after preliminary trials, which included a number of failures and premature detonations, it consisted of a normal spherical mine from the top of which a copper antenna was stretched up by a buoyant float. The mine operated on the simple principle of contact between the steel hull of a U-boat or ship with the antenna wire through a copper element inside the mine, creating a sea cell which caused a battery to fire the electric detonator.

With the antenna float roughly 10ft beneath the surface the mine itself was moored at a depth of some 50ft, thus presenting with its antenna a danger range of 40ft in depth for any passing submarine. Originally a lower antenna doubling the effective danger depth was used, but difficulties experienced in laying without damaging the parts, added to other technical problems, caused the double antenna mine to be scrapped. In the later mines electric switch horns were fitted round the mine case. These sensitive horns were designed to catch a submarine which might brush against the mine without touching the antenna above it.

It was agreed that British minelayers should take care of both the Scapa Flow end and the Norwegian part of the barrage, laying the standard British horned mine, but that for the middle section of the barrage, roughly 160 miles over the deepest part, the Americans would manufacture all the antenna mines and lay them.

For this operation ten American minelayers made their base at Cromarty on the Moray Firth, and between June and August 1918 they laid no fewer than 56,000 mines. Over the two end sections of the barrage five units of the British minelaying fleet laid a further 15,000 contact mines. The delivery of such numbers of antenna mines from factories in the USA across the Atlantic in convoy to anchorages in the Western Isles was no mean feat; the journey by rail from the Kyle of Lochalsh across Scotland to Cromarty was a major operation in transport, and for a time strained the single track working of the then Highland Railway to its limit.

Indeed the planning and laying of the great northern mine barrage was in itself one of the largest combined operations of the First World War, and the American allies were justly proud of the major part their manufacturers and Navy had played in completing the job in record time. But because of official secrecy and the serious news arriving from other theatres of the war, little was known about the event by the general public.

The fact remains, however, that this mine barrage had an immediate effect on the operations of U-boats, which found it more and more hazardous to break out of the North Sea into the convoy routes in the Atlantic, and the dire threat of starvation of the British people and the cutting off of vital war supplies was narrowly avoided.

Within three months of the completion of the great barrage, however, in November 1918 the Armistice was signed, the four years of the Kaiser war came to an end, and peace celebrations filled peoples' minds. But for the Allied navies problems still remained: all the mines sown during the war – not only those seventy thousand or more in the northern barrage but all those others laid around the British Isles and off the coasts of Germany – now had to be swept, and channels cleared for peacetime shipping. It was to be a vast clearing up operation, and the methods used deserve a chapter to themselves.

Chapter four

SHIPS HAVE THEIR SIGNATURES

Altogether during the First World War nearly a quarter of a million mines were laid by the countries involved and by neutrals protecting their own harbours. Of this number Britain was responsible for about 129,000, nearly half of them around enemy coasts and in the Mediterranean. The US Navy, as we have seen, laid in record time 56,000 mines in their main section of the northern barrage, together with some smaller fields in the northern approaches to the Irish Sea. Germany's minelaying accounted for 44,000 mines at points scattered around the British Isles, in the Baltic, the Mediterranean and the Black Sea. That mining had become a principal factor in sea warfare was demonstrated by the fact that through mines alone the Allies lost a total of a million tons of shipping in 586 ships, while German losses through the same cause were 148 warships and 36 U-boats.

Now all these minefields had to be removed and the seas cleared for peacetime shipping as rapidly as was practicable – and with the existing minesweeping methods this was going to be a long and anything but easy task, for mines moored in extensive fields and ready to fire are prickly objects to deal with, and not to be treated in undue haste. There had as yet not been much development in the techniques of minesweeping, and it was still something of a dreary and plodding job.

In the early days of mining the operation of clearing an area was excessively slow and primitive. The usual method was by a loop of ground chain which was towed from two spars set athwart a small vessel's stern so as to extend the width of the loop. Where mines of the electrically controlled type were to be swept, hooks and grapnels were added to the chain to sever the cables leading to the shore.

With two vessels towing a loop of wire or chain between them a wider sweep was possible, but it was never easy for two towing vessels to keep station without either closing the loop with the consequent width of sweep, or widening the gap and stretching the sweep almost up to the surface, so perhaps skimming harmlessly over the tops of the mines. In either case the ground sweep could come fast on a rock or a foul bottom, and part. This was a vexation that occurred only too often and helped to make the work of the sweepers an unhappy one. There was also the ever-present risk of one or other of the towing vessels being blown up as they worked their way steadily through the minefield.

Then someone noticed how the North Sea fishermen kept the mouths of their trawls stretched open by means of a pair of otter-boards, or doors, which were made to sheer away in opposite directions while being towed through the water, on the principle of a flat kite. A wire sweep was accordingly tried with a pair of otter boards to give the loop the desired spread; but the old troubles with a foul sea bed and parted sweeps were still encountered. For a time these snags were accepted and nothing much was done to improve the rate at which minesweeping could be carried out.

The principle of the otter-board was seized on, however, early in the war as a protection device for warships. This was the *paravane*, which consisted of two wires leading from the vessel's bow below the waterline (the forefoot) to port and starboard respectively. With the ship underway, each wire was kept at an angle of about 45 degrees from the bow by an otter, or paravane, at its outer end, designed so as to keep at its proper depth. The wires were serrated, that is a single strand of steel wire was laid counterwise to the lay of the rope. This presented a rough outer surface which would tend to saw through any mine mooring rope as it ran along the wire away from the ship's bow. A saw-tooth wire cutter near the outer end also made sure that no mine mooring remained uncut. Developed in various details, the paravane became a necessary adjunct to all operational units of the fleet, and was later fitted to merchant ships, until influence or non-contact mines were introduced and an entirely new form of ship protection had to be devised.

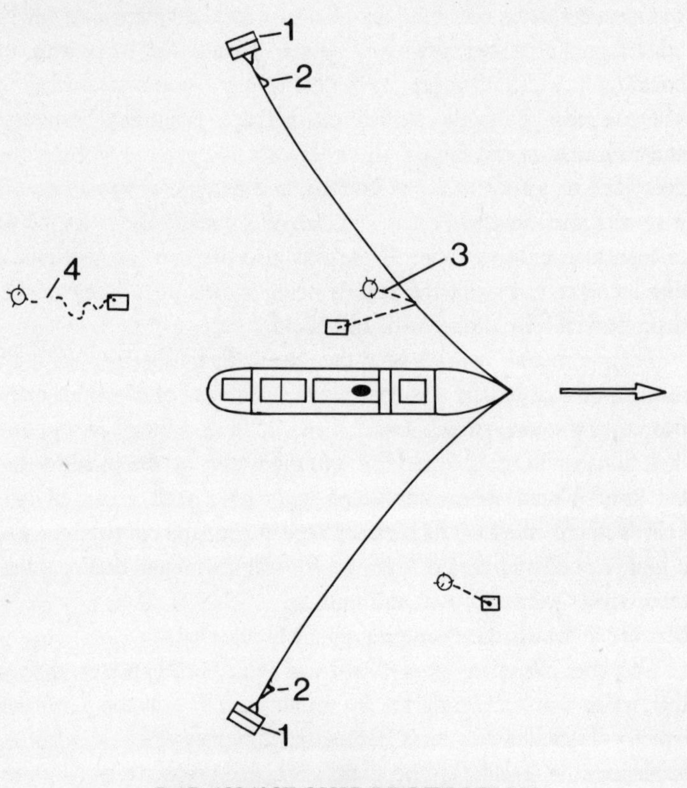

PARAVANE SHIP PROTECTION
1. *Otter*
2. *Wire cutter*
3. *Wire catches mine mooring*
4. *Mine cut*

The pressing need to clear not only all the mines in the northern barrage but the numerous minefields in both British and enemy waters following the Armistice demanded new sweeping techniques with a faster and better performance. When used with an otter at the end, the earlier sweep wires tended to take up a great downward bow between the stern of the sweeper and the otter, and thus varied in depth from end to end according to the speed of towing. Something was needed to keep the whole length of the sweep wire flatter, in other words at a more uniform depth from the vessel's stern to the otter.

Once more *Vernon*'s experimental teams came up with a solution and produced early in 1919 the *Oropesa* sweep, so called from the fleet sweeper with which the trials were carried out. In this sweep the wire was towed astern of the minesweeper with a multi-plane otter, or kite, attached to the wire a short distance from the point of tow, thus keeping this end of the wire at the required depth. Near the outer end an otter forced the wire away from the vessel, keeping it as taut as practicable, while attached to the otter by a short line was a torpedo-like float which ran on the surface, thus holding the otter from diving towards the bottom. Specially designed high-speed wire cutters were fitted at intervals along the wire, so that any mine moorings coming into contact with the wire would at once be cut, and the mine disposed of by gunfire.

The Oropesa sweep, which was worked by a single vessel, was in many ways easier to handle than the earlier loop sweep towed between a pair of sweepers, and reduced not only the time of clearing an area, but of course used fewer sweepers. Another advantage was that where sweeping was being carried out by a flotilla, only the leading vessel took the calculated risk of steaming along the edge of a virgin minefield; all the other sweepers following in formation were protected by their leader's sweep.

Excellent though it was in the vast areas to be cleared, the Oropesa was not suited to a requirement that arose for a sweep which could be carried out ahead of the faster units of the fleet. What emerged for this purpose, and was referred to as the High Speed Sweep, was a cross between the paravane and the Oropesa and was suitable for towing by destroyers.

This sweep comprised a bridle from the destroyer's stern having a specially modified multi-plane kite which kept the outer end of the bridle at the required depth. From the bridle, just ahead of the kite, sweep wires led off to port and starboard with a paravane at the end of each. Viewed in plan, the whole sweep formed roughly the letter

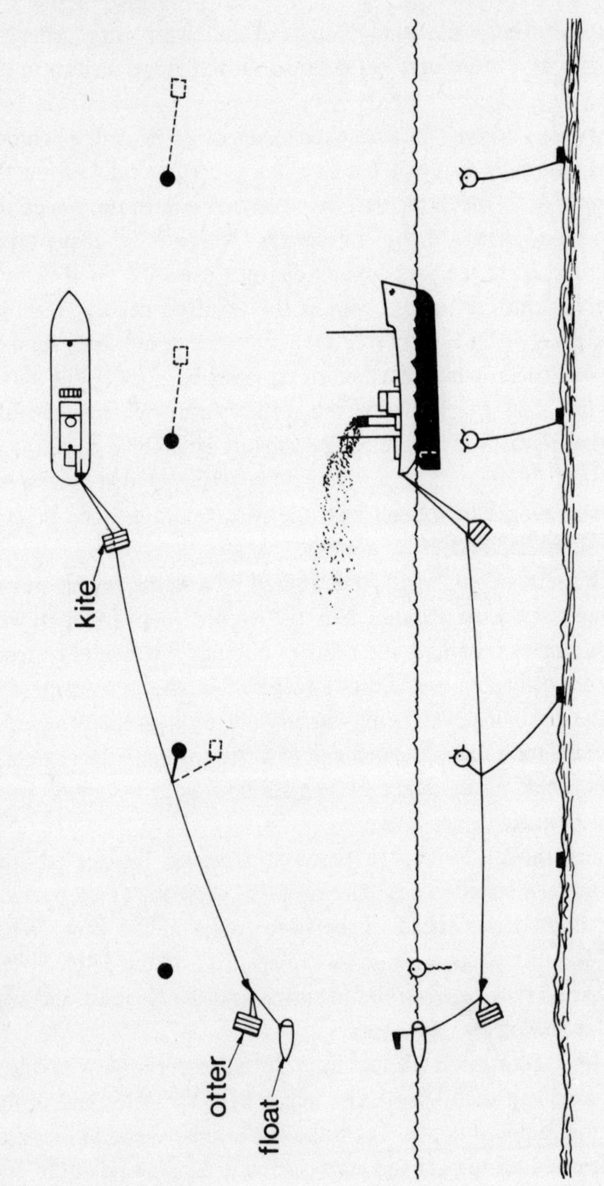

OROPESA SWEEP

Y. It proved capable of skimming a comparatively wide channel of some 400yds at speeds around 15kts. Inevitably, owing to the enormous strain on the gear and for other technical reasons, the width of sweep became narrower as the speed of tow increased, and at 25kts the effective width was reduced to some 50yds only. Even this was considered acceptable for a single following ship when circumstances justified the risk.

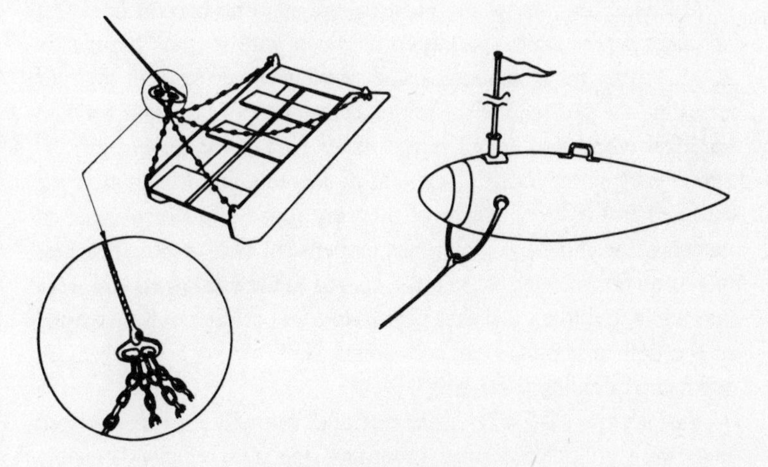

ITEMS OF SWEEPING GEAR
Oropesa float (right) and otter (kite), with inset showing detail of the link coupling between towing wire and otter chain

By international agreement mines which break adrift from their moorings should automatically render themselves safe. This was usually achieved by a spring device which would neutralise the detonator as soon as the pull on the mooring wire was released or cut any electrical circuit that might be involved. After mines had been in the sea for a considerable time, however, such painstaking fittings were liable to become corroded and unreliable. This was probably one of the reasons why so many mines, parted from their moorings by a storm, would blow up when they reached the shore and were rolled over by the waves on to one of the horns.

Apart from laudable agreements between nations, any manoeuvre in war calls forth appropriate counter-measures. It is to be expected, therefore, that whichever opponent laid mines would endeavour to make the task of clearing them as difficult as possible. All sorts of anti-sweeping devices were accordingly proposed and tried to prevent mine moorings from being cut.

Mechanically, one of the most ingenious to be introduced was a sprocket wheel which permitted a sweep wire to pass harmlessly *through* the mine mooring rope without severing the rope or revealing the presence of the mine. The spindle of this eight-toothed sprocket wheel was linked to the upper part of the mine rope, the sprockets running inside a segment of curved guide which was itself attached to the lower length of mooring rope. The sweep wire, on catching the mooring cable, ran in between two of the sprockets which turned and led it over the curved guide and away. The idea was attributed to an assistant paymaster, and it was wryly suggested at the time that only a paybob would think out such an ingenious method of evading something!

Various types of snags, grapnels and wire cutters with serrated jaws were attached to mine moorings, the most efficient possibly being an explosive cutter whose jaws were rammed shut by a small powder charge triggered off by the sweep wire. The Germans concentrated on a highly effective form of sweep obstructor like a dummy mine, in the form of a float on a line and sinker fitted with an explosive cutter beneath the float. It was their policy in some layings to plant at least one of these cutters for every moored mine, and on occasion there were many more cutters than mines.

This kind of mixed field was undoubtedly an embarrassment to the sweepers, whose sweeps were being constantly cut and lost without destroying any of the mines. But by dogged perseverence – with which the British are said to be highly endowed – it was just a matter of time before all the cutters were used up and the mines themselves could be swept in the usual way. This prodigal strewing of explosive cutters is only a matter of policy, and the view has been expressed that wire cutters don't destroy ships, and the enemy might have claimed better results by replacing them with mines.

Another crafty method of foxing an enemy's minesweepers, which was resorted to by all belligerents, was to lay some of the mines with delayed release sinkers. At some time after the mines with normal sinkers had been discovered and the area swept and declared clear, a number of other mines would be released from their sinkers and hopefully sink an enemy ship or two. This could be repeated later on and a third group of mines rise from the sea bed and present another hazard. With this device the mine on being laid sank to the bottom with its sinker and remained attached to it by a link or short wire seizing, until either a soluble plug, or for more accurate timing an electric clock, caused an explosive cutter to sever the connection and allow the mine to rise. A preset hydrostatic device in the sinker controlled the amount of wire that was to unwind from its drum, so that the mine was held at the required depth setting.

The most elusive type of mine to sweep or recover is the one that rests on the sea bed. With all heavy objects, from a mine to the wreck of a ship lying on the bottom in a tideway, the effect of the tidal streams is to scour a kind of moat in the mud and sand around the object. With each flood tide flowing past in one direction, and then the ebb running the other way, the sand is scoured more and more while the heavy object, little by little, sinks deeper inside the pool formed around it. Depending on the nature of the sea bed and the strength of the tides, the wreck of a ship can become half submerged in the sand within a month or so, while a ground mine has been known to be completely covered over inside a fortnight. Any attempt to sweep or destroy such a mine calls for entirely different techniques.

It has already been seen that the British were ahead of the Germans in evolving and using a magnetically actuated ground mine towards the end of the First World War. Experiments in firing a mine solely by the sounds made by a ship through the water – in short, an acoustic mine – were also well advanced by August 1918, and both magnetic and acoustic units were at first intended to be used in mines of the moored variety.

But there was as yet no known way of sweeping such mines with safety, for the sweepers themselves would be in danger of exploding these mines while steaming over them. Although the devastating effects on a ship from a mine on the sea bed were yet to be realized, the Admiralty decided to lay these first magnetic mines on the bottom in shallow water; as they would not be in contact with any ship, but probably several fathoms beneath the hull, a powerful charge of 1000lb of TNT was used, with its mechanism set within a heavy concrete base which would not be moved by the strong tides off the Belgian coast.

This mechanism was of a simple type in which a pair of compass needles moved when a ship passed overhead, and the change of angle, or dip, between the needles made electrical contact and exploded the mine. In all, 400 of these so-called British M-Sinkers were laid off Zeebrugge, and a further forty outside Ostende. A few blew up prematurely, but others were known to have sunk a German destroyer and at least one other enemy craft, while presumably the remainder still lurk – although harmless enough now – in their resting places. As unsweepable problems, they had been fitted with time-expired self-flooding devices.

A magnetic mine works on the principle that every ship built of steel carries some amount of permanent magnetism. A steel ship is, in effect, a large if somewhat weak magnet, with a north-seeking pole at one end and a south-seeking pole at the other. Which pole is at, say, the bow end depends on a variety of factors, among them whether the ship was built in the Northern Hemisphere or the Southern, and which way her building berth lay in relation to the North or South Magnetic Pole, in other words the compass bearing.

A part of this so-called permanent magnetism is induced during construction by the hammering and riveting and general vibration to which all parts of a ship's hull are subjected. Every ship therefore has her own degree of induced magnetism, and if she is passed over a special instrument this can be recorded in the form of a graph. Every ship, however, will record a different curve, and like a person's fingerprints this wavy line can be identified with that one ship alone. It is appropriately called the ship's Magnetic Signature.

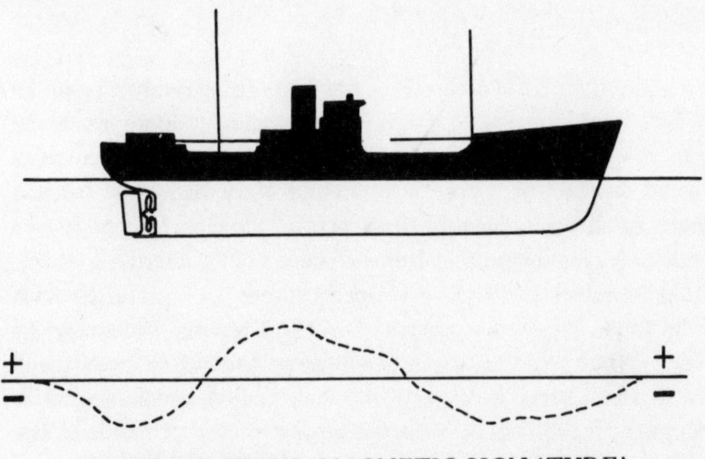

TYPICAL SHIP'S 'MAGNETIC SIGNATURE'

This may be an oversimplified description of a ship's magnetic state, and we will consider it in more detail when we come to the question of how a ship can be immunised against magnetic mines, but for the present it is sufficient to appreciate how both the British and the Germans seized on this natural phenomenon as a well-nigh perfect weapon of ship destruction.

Chapter five

STRATEGY – BY THE THOUSAND

To imply that at the outbreak of war again in September 1939 the British were unprepared for full scale mining warfare would be grossly wide of the mark. Already in store at home and in bases spread around the Commonwealth the Navy had some 20,000 mines of all types. Most of these were of the moored contact type with both plummet and hydrostatic sinkers of an improved design.

These mines had been developed between the wars to be fitted with 500lb and 330lb charges of TNT. The mine containing the larger charge could be laid in depths up to 500 fathoms, while with the smaller charge which gave the mine case the additional lift or buoyancy needed to hold up the greater weight of mooring wire against the pull of the tides, the mine could be laid in depths up to 1000 fathoms, and, dependent on the type of sinker fitted these mines (officially called the H II type), could be set to be held at any depth below the surface to 75 fathoms, for use against either surface ships or deep running submarines.

In addition, a new type of magnetic ground mine had been developed with a substantial 880lb TNT charge. The mine case was in the form of a cylinder with one end shaped so that in sinking the mine would, in theory, always come to rest lying flat on the sea bed. The overall measurements were also planned for the mine to be fired from a submarine's torpedo tubes, while a similar type designed for dropping from a torpedo-carrying aircraft, or from specially adapted bomber's bomb racks, was already in production.

The original simple firing method used in the British M-Sinkers in 1918, on the dip needle principle, had been abandoned by 1932 and a greatly improved system incorporated in these new mines. Briefly, this comprised a solenoid consisting of fine copper wire wound round a nickle iron and copper rod which ran almost the whole length of the charge case. The solenoid transmitted the current, induced in its coil by a ship passing close enough, to a highly sensitive relay connected to the firing mechanism.

The principle of actuation differed from that employed in the old M-Sinker, which had operated on the sharp increase of magnetic field induced by the ship, in that the new mine reacted solely to a rate of *change* in the magnetic field, from north to south or vice versa. This was an important improvement which was to make the mines difficult, although not impossible, to sweep.

In the use of all this accumulated material, the tentative approach to minelaying which had been such a negative feature in Whitehall at the outbreak of the First World War was now changed into a vigorous policy of mine warfare. Standing by and already loaded with mines were a cruiser, two destroyers converted for minelaying and the specially constructed 800-ton minelayer *Plover*, known as the Vernon Pet*. Six minelaying submarines were available at short notice, and other vessels already being converted to carry larger numbers of mines included some forty more destroyers, two train ferries, a car ferry and a merchant ship.

With the experience gained of the enemy's preliminary tactics in the First World War, plans had been formulated many months in advance of the minelaying operations that had to be put in hand as soon as war was declared. When the moment came speed was vital to the element of surprise, and in sorties that lasted only five days four of the minelayers completed the laying of 3,000 mines at both deep and shallow settings in a line stretching from the southern end of the Goodwin Sands to the Dyck Bank off Dunkirk. The whole operation was finished by 16 September 1939, while the French Navy assisted by planting their mines in the remaining gap between the Bank and the inshore channels.

There were signs in *communiqués* which later came to light implying that the speed with which this operation had been carried out had taken the Germans by surprise. They did, however, expect a

* *Plover* was broken up in 1974, at 38 years the oldest vessel in active service in the Royal Navy.

repeat of the first war's mine barrage to be laid across Dover Strait from Folkestone to Cap Gris Nez, and this mixture of moored contact mines and mined submarine nets was duly completed by the two converted train ferries by the end of September. Surprisingly enough the work went on without any harassment from the enemy, but as events turned out with the occupation of France the value of both barrages as deterrents was lost at the Continental end of the line. Nevertheless they were known to have sunk at least three U-boats in the first six weeks.

Continuing for the moment this review of minelaying policy adopted by Britain, extensive minefields of moored mines were laid during October and November 1939 up and down the East Coast so as to protect shipping off the coasts of England and Scotland from expected attacks by U-boats, and the whole area, as required by International Law, was declared to be dangerous.

Because it needed enormous numbers of mines as well as time to cover such a large area, a number of *dummy* mines were laid down the middle of the planned lines until this East Coast barrage was completed. These 'non-mines' were laid with their moorings carefully adjusted so that the mines would be awash – or 'watching' – for an hour or two each side of low water. It was hoped that if sighted in this state by enemy reconnaissance aircraft or surface patrols they would give the appearance of an extensive minefield, and as subsequent reports indicated this artful ploy did help to keep German activities to a low profile for a time, and allowed coastal convoys a less hazardous passage.

Whereas the attacks on the Atlantic convoys by U-boats almost brought Britain to her knees in 1917, until the combined British-American laying of the Northern Barrage from the Isles of Orkney to the coast of Norway helped to contain most of the enemy submarines in their home waters, the Admiralty knew full well that this time German U-boats would be an even greater threat to the Allies' shipping. Plans were well advanced, therefore, to repeat the northern barrage scheme with the highest priority.

Although the vast resources of American minelaying would not now be available, the Navy had built up the necessary stocks of mining material, new minelayers were coming into service suitable for these stormy waters, and factories up and down the country were working at high pressure to produce the mines, sinkers and all the necessary gear. The old antenna mine had been updated in the years before the war and offered a far greater danger depth for both surface and underwater vessels, and this was the type chosen for the purpose.

The question where to have the shore bases for the huge amount of materials involved was not a simple one. As the previous American bases in the Moray Firth would in this war be liable to intensive enemy air attack, convenient though the north-east coast was for the barrage, the Kyle of Lochalsh, facing the Western Isles, was the place selected. This base had already had its importance, as we have seen, during the first war, for it was here that the American-produced antenna mines and their gear were landed for use in the old Northern Barrage, and transported by rail to the bases on the East Coast.

Now, such are the reversals in the fortunes of war, all the materials needed for the new barrage had to go in the *opposite* direction; from the munitions factories they were sent by trainloads via Inverness and over the hard-pressed single line which was no longer the Highland Railway, but a tortuous, heavily graded branch of the London, Midland and Scottish system. But before the first minelayers could be loaded and routed round Scotland and through the Pentland Firth into the North Sea, the whole situation was changed.

Hitler had brought off another of his inspired coups by occupying Norway – almost, it seemed, overnight – and the value of any lines of mines across the North Sea was rendered void. It was now impossible to lay mines up to the occupied coast, nor could the eastern end be adequately patrolled. The old Northern Barrage scheme had to be abandoned.

Yet it was imperative that some form of obstruction should be laid to deter the fleets of U-boats from slipping out round Scotland and continuing their devastating attacks on the Atlantic convoys. The alternative scheme which was put in hand at once was even bolder. It was to lay a series of minefields on the routes that the U-boats were believed to take on their way into the Atlantic, between Cape Wrath on the north-west tip of Scotland and the east coast of Iceland.

On the face of it this was a daunting proposition and at once raised new problems. For one thing, the distance to be protected, roughly 480 miles, was nearly twice as great as the old North Sea barrage (250 miles), depths in between were greater and reached as much as 500 fathoms in places, and ocean currents, such as the North Atlantic Drift, were stronger than North Sea tides and would tend to drag the long lengths of mine ropes and their antennae into great bows, causing the mines to dip well below their correct settings. Furthermore, worse weather conditions and heavier seas were prevalent in this area compared with those to be found in the northern North Sea.

With the estimate of 65,000 mines required for the original barrage scheme ready and stored in lighters and railway depots around Loch Alsh, the Admiralty decided that this operation should go ahead forthwith. This north-western barrage, however, needed far more mines, and production was again stepped up and priority given to their delivery to the Kyles. Even so, there was no question of being able to lay rows of mines at close intervals all the way from Scotland to Iceland; such a barrage would take more than the whole country's production capacity of mining materials.

The line, started off Cape Wrath, could not be continued progessively towards Iceland, for by the time the end of the line was reached the first mines would have come to the end of their active life. Moored mines and their mooring wires and firing mechanisms do not have an indefinite life when in the sea. The mines were therefore laid in groups, but to give credence to the suggestion that there were no considerable gaps, the entire area was announced, in accordance with International Law, as dangerous, and neutral shipping was given safe courses to be followed through the Faeroes or the Pentland Firth.

Shortly afterwards, similar scattered minefields were laid across the Denmark Strait between Iceland and the edge of the Greenland ice pack. This was another stretch of some 260 miles and brought the western end of the barrage at least 700 miles steaming distance from the minelayers' base at Loch Alsh. For this Strait, which is subject to drifting ice from October through the winter, moored magnetic mines were used in place of the antenna type. The reason for this was that magnetic mines can be set deep enough to avoid being carried away by icefloes, while as non-contact mines they can still severely damage, if not sink, either a submarine or a surface vessel.

Altogether in these Scotland – Iceland – Greenland fields ships of the First Minelaying Squadron laid some 96,000 mines of both the antenna and the moored magnetic type, a truly remarkable effort in the space of five months. There can be little doubt that the expenditure of material and time and human endurance in laying this system of grouped minefields was justified, for their very presence in positions unknown in detail to the enemy imposed a limitation on the movements of both U-boats and surface ships.

To this extent these minefields made an effective contribution towards protecting the numbers of convoys which were at all times making their way across the Atlantic, for it was recorded that seven U-boats were lost in this area which were not sunk by depth charge attacks. It was also the presence of these minefields that led to the discovery of enemy forces in the area and to one of the most spectacular chases in naval history, for had not *Bismarck* and *Prinz Eugen* been forced to use the safe channel between the edge of the pack ice stretching out from the Greenland shore and the minefields along the west coast of Iceland, the German ships would have stood a much better chance of slipping through and out into the Atlantic without being spotted in the prevailing thick weather. As a battle was said to have been lost through lack of a horse shoe, the *Bismarck* was steered to her own destruction – without sinking a single merchantman – through the presence of dormant mines.

The reverse side of the coin, however, showing the complete

indifference that mines can have for both friend and foe, was demonstrated by an unfortunate incident affecting one of the Arctic convoys which was returning from Northern Russia the following year, 1942. On orders from the Admiralty the large convoy was split up into two parts when approaching Iceland, one section being routed down the east coast on its way to Loch Ewe in north-west Scotland. The other section, comprising nineteen merchant ships with three naval vessels and two armed trawlers as escorts, was to pass to the westward throught the Denmark Strait and so to Hvalfjord on the Icelandic south-west coast.

After steaming for days through thick weather the position of this convoy in relation to the narrow safe passage in the Strait which had proved fatal to the *Bismarck* was very doubtful. The leader of the escorts, the minesweeper *Niger*, accordingly forged ahead into the mist so as to get a sight of the land and fix the convoy's position. Unhappily, the loom of an icefield was mistaken for the shoreline, and the ships of the convoy were ordered to alter to a course which brought them on to a bearing straight into the minefield.

The *Niger* was mined and sank in a few minutes, and in succession five of the freighters were blown up and lost through British-laid mines, while a sixth merchantman was badly damaged, but her crew managed to keep her afloat and on the move. While the two escort trawlers gallantly stood by in the midst of the mines to pick up survivors, one of the minesweepers steamed farther in towards the coast and managed to get a definite fix on their position, and the remaining vessels were led safely back into the clear channel.

These incidents have clearly shown, it is hoped, the two principal uses of minefields: first, for the *protection* of harbours, naval bases, convoy anchorages and friendly or neutral shipping, and secondly, for the *strategic* deployment of mines to influence or restrict the movement of enemy ships, or even to lure them towards a situation where they might later be destroyed.

If ever a competition was needed to demonstrate the relative effectiveness in ship destruction of big guns, torpedoes, aerial bombs and mines, one was held in the English Channel early in 1942. The incident created a furore in sections of the British press, which described it as the worst humiliation suffered by British naval power since the Dutch raided the Medway in 1667. It was, of course, the famous 'Channel Dash', in which the German battlecruisers *Scharnhorst* and *Gneisenau* and the heavy cruiser *Prinz Eugen* made their bold attempt to return to Germany from their French base at Brest.

On the face of it, it appeared that neither guns, nor torpedoes, nor bombs had had any effect whatever on these ships' successful voyage home. But although they had made their intended ports, they had not entirely escaped the hidden menace, for twice during the afternoon and evening of the previous day the *Scharnhorst* had been damaged by mines. And the same evening the *Gneisenau* had also been mined, and so seriously was she damaged that she was never again in service. These were some of the mines which had been laid by RAF Bomber Command, and they had proved in the end the most effective deterrent.

It must not be inferred from the above, however, that the Admiralty had been lax in preparing for the expected sortie by the Germans. Anticipating their probable course if they came up Channel, the two fast minelayers *Welshman* and *Manxman* weeks before had carried out the laying of twelve separate minefields off the French coast. To hamper the German sweepers as much as possible they had laid both contact and moored magnetic mines interspersed with a percentage of delayed release mines. And they had completed the whole operation in fourteen consecutive nights: a fine instance of hardworking perseverence close to the enemy's coast.

Part of the Dover mine barrage had also been reinforced with moored magnetic mines by *Plover*, the *Vernon* minelayer, whilst aircraft of Bomber Command had carried out minelaying sorties, in daylight with only moderate cloud cover so as to ensure greater accuracy in laying, and dropped groups of magnetic mines off the Frisian Islands and in the approaches to the Germans' principal naval ports.

All these minefields played some part in disturbing the enemy, as in addition to the two battlecruisers a destroyer on her way to augment the destroyer escorts at Brest was struck – fatally – by a mine of the Dover barrage, while another of the escorts was mined and sunk off the French coast.

As a different type of weapon which is difficult to evade but can be deadly in its effects, and was used with success on many occasions during the war, the limpet mine should take its place in history. Related as it is to the very earliest type of underwater mine – the 'infernal machine for attaching to the bottoms of ships' used by both Bushnell and Fulton – the modern limpet was in fact developed before the war in 1939 in the MD1 laboratories of the Ministry of Defence.

As finally produced it comprised a circular container like a plastic clamshell, filled with from 4lb to 10lb of blasting gelatine or a similar explosive, and fitted with a time fuse. These mines were placed in position on the bottom plating of enemy ships by frogmen working from a midget submarine or from some hidden vessel. Flexibly mounted magnets on the rim of the mine held it to the steel plating however rough the surface of the hull might be, and the magnets were powerful enough to keep the mine in place even if the ship later got underway and steamed at 20kts.

The time fuse could be set for a period from 30 minutes to anything up to five fours. Many daring raids were made by frogmen with limpet mines by both sides during the war and were the cause of considerable damage to ships. Even if the vessels were not always sunk the repairs necessary took valuable time in the shipyards, while in many cases the moral effect on ships' crews was greater than the damage caused. As a form of silent attack with worthwhile results it can be recorded that during the six years of the war the Ministry of Defence authorised the issue of more than half a million limpet mines for use against enemy ships and installations.

It is time, therefore, to examine what part mines played in German strategy, and the impact the first of Hitler's secret weapons had on the British war effort.

Chapter six

A SECRET IS UNCOVERED

As October merged into November of that first year of the war the numbers of ships sunk each day in the channels leading into the Thames Estuary became even more grave. In a room occupied by Commander M (Mining) in *Vernon*, a large-scale Admiralty chart on the wall showed by its growing clusters of pinpoints and ships' names the casualties which were recorded each morning by a Lieutenant RNVR, all of them caused by mines of an influence type.

But what type? It had long been known that if war came again the Germans would use magnetic mines, as the British had done towards the end of the 1914–18 conflict. But the Germans could also be using sound-operated or even pressure mines, for both these types had been tried experimentally by the inventive British some months before the Armistice.

While every effort was being made to discover effective methods of sweeping these mines, as yet there had been no results; the mines refused to respond to any of the normal sweeping runs, except beneath some unfortunate passing ship. Until a specimen was obtained and its method of working studied, therefore, nothing could be done to prevent more and more vessels being destroyed – except to hold up all ship movements off the East Coast pending a solution to the problem. But with so many cargoes desperately needed, that negative course was unthinkable. Hitler had promised to use a secret weapon, and it looked from all angles as if this was it – and there was as yet no answer to it.

All the evidence so far had indicated that the mines were being laid by aircraft flying low at night when the moon was nearly full, and that they were not of the buoyant moored variety but mines that settled on the bottom. Someone then asked, why not try recovering one of them by dragging a strong sweep or net along the sea bed, then hauling the 'catch' ashore somewhere safe, where the object could be dissected and its secrets made known?

It didn't seem such a crazy idea at the time, since all other proposals had failed. And it so happened that one of the *Vernon* officers, Commander C E Hamond RN, who had returned to the Service after some years of retirement, had first-hand knowledge of trawling and understood the jargon of the North Sea trawlermen. Grey haired, bearded and with the bluff manner of an Elizabethan shipmaster, Hamond was the natural choice to be given the task of devising a suitable form of ground trawl gear which could be made entirely of non-magnetic materials. In due time, for such materials were in short supply and a strong net of bronze wire was no easy thing for landsmen to make, two non-magnetic trawls were available, together with a brace of North Sea steam trawlers which had been allocated for the purpose. Mines had been sighted dropping into the sea on dark-coloured parachutes in the area of the Tongue, in the Thames Estuary, and here the two trawlers, *Cape Spartel* and *Mastiff*, began their work with Lieutenant J Glenny RN in charge of the operation.

The second vessel's trawl came fast on what appeared to be a mine, and while she began to haul in the trawl warps with the steam winch the *Cape Spartel* stood by. It is as well she did so, for a short distance from the *Mastiff*'s stern the mine detonated with a heavy roar and a mountain of spray momentarily hid the little trawler. When it cleared little was to be seen of the *Mastiff* except her funnel and fo'c'sle disappearing beneath the surface and a wrecked Carley float. Prompt action by the skipper of the *Cape Spartel* managed to save most of the *Mastiff*'s crew who had been thrown into the water, but five men went down with their ship.

This was a serious setback for the advocates of retrieving a mine by trawling, and the operation might have ended right there; but on reflection it was decided that as there appeared at present no other way of recovering one of these mines, the urgency of the matter justified the risk. One thing, however, had been made very clear: the steam trawlers with their steel hulls and deep draught were totally unsuitable for working over the shallow waters in which these mines were being dropped, and only vessels with wooden hulls having the lowest magnetic field could be used with any degree of safety.

This implied a choice of some of the older and smaller wooden *drifters* which could be converted for trawling. The Navy already had a number of these handy little steamers – motor drifters had not yet come on the scene – but they were all allocated for ferrying men and stores between ships and base, laying dan buoys for the bigger sweepers and other inshore duties, and could not be released. With Lieutenant R S Armitage RNVR as his Number One, Commander Hamond accordingly obtained powers to requisition up to a dozen of these wooden steam drifters which were to be found lying idle and unable to fish at various ports.

In the midst of the operation the author was summoned to the Commander's office in the Portsmouth base.

'Come on, Griff,' rumbled the quarterdeck voice, 'you're supposed to know something about boats, and steam drifters in particular. There are a couple of drifters lying at Aberdeen, and two more at Buckie and Peterhead, all wooden ones and pretty long in the tooth, I shouldn't wonder. But they might suit us, so get a railway warrant and go and have a look over them. See if they're in good enough condition to take our non-magnetic trawl gear without pulling themselves apart. Then report back to me.'

Welcoming an opportunity to see something of Scottish fishing ports which were new to him, Griffiths clambered all over the weather-battered and mouldy little tubs which reeked of fish, and pronounced three of them serviceable for conversion to naval work, doubtless to the immense relief of their canny owners. The fourth at Buckie had engine trouble and a leaky boiler which only the oldtime steam engineer's remedy of a peck of dried peas might have cured for the time being, and like so many old wooden boats she was too far gone, even for expendable duties.

And so, together with other drifters culled from Fraserburgh, Lowestoft and Great Yarmouth, these vessels were rapidly converted to special trawling, and became the first of what were known as

Vernon's private navies. It is perhaps not out of place to explain that the essential difference between trawlers and drifters is in their respective methods of fishing and the way they work their gear. The trawler tows her net astern of her, with its mouth stretched open by a pair of otter boards, each of which is attached to a steel cable or warp which is led, one to a gallows, or A-frame, near the stern and the other to a similar gallows near the bow. Both warps are thence taken through roller deck sheaves to the two separate drums of a steam winch mounted on deck between the fish hold and the wheelhouse. The two gallows frames ('galluses') on both sides of the vessel, to allow the trawl to be shot either to starboard or to port, together with the big steam winch amidships are unmistakable features of the offshore trawler.

The herring drifter, on the other hand, rides by the bows to her nets, which stream out ahead of her like an underwater curtain as much as two miles in length and from six to seven yards in depth. Weighted along the base with lead pellets, the nets are supported at the surface by a row of blobs, or floats of glass or plastic balls. Shoals of fish get their gills caught in the fine mesh, and when hauling begins the nets are brought in round a chattering steam capstan on the foredeck. While the men disentangle the silver creatures and drop them into the fish hold, the deck boy flakes the incoming net down in the net locker below. Drifters riding to their nets with foremast lowered and a smoke-blackened mizzen sheeted in over the stern to help keep them head to wind have long been a familiar peacetime picture in the North Sea.

For conversion to trawling for the magnetic ground mine the capstan with its compact Elliott & Garood engine mounted on it was taken off, gallows frames and deck leads substituted, and a trawl winch fitted before the wheelhouse. The fish hold was thoroughly fumigated and cleaned out – so they said – and fitted out as a mess deck for the Hostilities Only crew, a bridge was built over the wheelhouse for the unit officer to control the sweeping operations, and Lewis machine guns were added as a morale booster. Painted grey all over and given an official number on the funnel, the old fishing vessels emerged as trim seekers of the invisible.

Unlike the bigger and more powerful trawlers hailing from Hull and Grimsby, Aberdeen and Fleetwood, these ex-drifters with their small compound engines were found to lack the strength to drag a non-magnetic wire trawl when the end became choked with mud, and when a mine weighing a ton or so did get into the net the whole outfit would come fast as though anchored to a rock. They had to work, therefore, in pairs with the trawl stretched between them as in the days of the old A-sweep. For this end each mine recovery group was accordingly made up of four boats in the charge of a Lieutenant RNVR as the unit officer, under the orders of Commander Hamond.

Transferring their unique custom from one part of the coast to another, ranging indeed from Falmouth in the West Country round by the Forelands and the east coast to the Scottish ports, wherever magnetic mines were reported, these *Vernon* units played their various roles with their own adventures and tribulations, their near-misses and minor triumphs. In October 1939 the approaches to the River Tay leading to the submarine base at Dundee had been mined, and a small Finnish steamer, bringing timber and foodstuffs from the Baltic, had been blown up just outside the bar. At least one other parachute mine had been spotted dropping into the main channel, and all ship movements were at a standstill in the port.

The local sweepers had been working over the area, so far without success – but at least these steel hulled trawlers with their sweep gear had not blown themselves up. Urgent problems sometimes call for desperate remedies, and although it was decided to permit small ships to pass well to the side of the channel away from the suspected position of the mine, it was of paramount importance that the mine should be recovered rather than destroyed.

As soon as the ships forming Mine Recovery Flotilla No 3 were ready for service early in November they were despatched in the charge of the author as unit officer, and set about working over the area with their non-magnetic trawling gear. On the third day of seeking, a heavy object was caught in the net being dragged between them by *Scotch Thistle* and *Achievable*. With excitement mounting on board each of the little ships, they slowly hauled the object out of the ship channel, shuddering under the maximum power of their engines, towards the sandy beach. But alas! in the shallows of Buddon Ness there was suddenly a mighty roar, a mountain of dirty water astern, and all hopes of a retrieved specimen had gone. Some hydrostatic switch had perhaps functioned correctly!

No more mines were located while the *Vernon* flotilla worked out of Dundee, and the area was in due course declared clear. For some reason the Germans left the Tay and its treacherous bar alone for the time being. Before receiving orders to proceed southwards along the coast in search of further minelayings Griffiths discovered one advantage in having a trawl: during a lull in mine search activities he remembered that when the drifters were being converted he had a normal fishing trawl included in the ships' equipment, in case it came in useful. And now the crew, most of them from east coast fishing ports, were given the opportunity for a drag with what they called 'a proper bloody trawl'. With fishing in the blood they had long bemoaned the fact that the mesh of the special non-magnetic wire trawl was too large to catch fish ('except a flippin' shark') and the resulting haul, when the trawl was hoisted at the jilson block on the foremast and spewed its slithery catch all over the deck, made up for all the days of hard but unprofitable work.

It may have been coincidence, but after a welcome supply of fresh fish had been distributed around the base, Griffiths found it far from difficult to draw any stores and gear his flotilla needed, while they were allocated from then onwards the most convenient berths in the Fish Dock for their mysterious comings and goings. It was, in fact, a happy situation, while it lasted.

In the meanwhile events were happening farther south in the mouth of the Thames which were to have far-reaching consequences. On the night of 21 November close to high water an unidentified aircraft was heard a mile or two to seaward of Southend-on-Sea pier. It was a calm night of almost full moon, but a low mist hung over the water and hid the beach from view. The plane appeared to circle twice, and then the sound of its engine died away.

Soon after dawn when the ebb tide had drawn all the water off the foreshore a black cylindrical object was sighted, lying partially embedded in the wet sand. A short distance away the sodden folds of a dark green parachute lay draped like some grotesque seaweed. The military authorities at nearby Shoebury Fort were alerted, and as the object was evidently some type of sea mine and below high water mark, the Naval Officer in Command (NOIC) Southend was informed.

At once a special *Vernon* party was despatched by car from Portsmouth with orders to get to Southend at all speed and to recover the mine 'at all costs'. And by the time they arrived at Shoeburyness a second object, similar in every detail to the first, had been discovered rather more deeply buried in the sand about 300yds nearer low water mark.

The *Vernon* party, in the charge of Lt Commander J G D Ouvry RN, with Lt Commander R C Lewis RN, chief petty officer Baldwin and AB seaman Vearncombe, began the highly ticklish task of opening up the first mine. It was found that a four pin spanner was needed to unscrew the metal keep ring of what looked like the primer pocket, and the ready help of the Army engineers in Shoebury Fort in promptly making a suitable tool in non-magnetic brass was deeply appreciated.

Step by cautious step, each operation carried out where possible from a distance, the mine was opened up; the shining metal primer, the size of a pint bottle, was drawn out by a long line, and at close quarters Ouvry unscrewed the metal gaine, or detonator, half the size of a man's thumb, and put it gently in a safe place.

When all the electric leads which had become exposed were cut, one by one, the object was declared safe to handle, and every one of the party breathed more freely again. The inert carcass with its more sensitive bits and pieces packed separately was loaded on to a Navy truck and hurried to *Vernon* for the experts to examine. This was the breakthrough the Admiralty had been so anxiously waiting for; at long last the secrets of what kind of mine the Germans were using, and exactly how its mechanism was activated would be laid bare.

To say that excitement in high places was intense would be a mere understatement, for Ouvry and Lewis were called to a meeting with the First Lord, Mr Winston Churchill, to describe every detail of the operation. In the meantime the second mine found nearby was rendered safe by another *Vernon* officer, Lt Commander J E M Glenny RN, following out Ouvry's detailed instructions. And while the two trophies lay in the *Vernon* mining shed under close guard, the King, George VI, paid Portsmouth a visit to see them for himself, when Ouvry, Lewis and Glenny were decorated with the Distinguished Service Order, and Baldwin and Vearncombe received each the DSM.

These first German magnetic mines had cylindrical cases of an aluminium alloy measuring 6ft in length and 22in in diameter. The hemispherical front end contained an explosive charge of 430kg (950lb), with six short open-ended tubes radiating the outside and designed to prevent by suction the mine being rolled over by the tide.

Inside the slightly coned rear section a strong coil spring was held in compression by a soluble plug. As the mine was dropped from the aircraft the 4yd diameter parachute was pulled out of this rear section, and enabled the mine to fall at a rate of about 60fs. The shock of striking the surface of the water was thereby controlled, but it was sufficient to start an arming clock running.

As the mine sank a spring-loaded button in the side of the case acted as a hydrostatic switch which, on reaching a depth of three fathoms pressed in a small spindle and so stopped the clock mechanism. After a few minutes, when the mine had settled on the bottom and the delicate mechanism had had time to come to rest, an electrical circuit was completed and the mine, hitherto inert, came 'alive' and ready to detonate under any suitable magnetic influence.

Meanwhile the soluble plug holding back the coil spring dissolved in the water, and the spring ejected the parachute. Thus it was devised that the tide should carry the telltale mass of green nylon folds and plaited cords well away from the mine so as not to give away its position. If, on the other hand, the mine had been dropped in less than three fathoms of water, or on to dry land, the hydrostatic button would not have operated and the arming clock would be allowed to tick away its preset time of 20 to 24 seconds, when the mine would detonate.

The mine was ingeniously thought out, therefore, to act either as a ship-destroying weapon on the sea bed, or as a formidable bomb or land-mine elsewhere. Furthermore, if attempts were made to recover this mine from the bottom of the sea, as soon as it was raised above the three-fathom depth mark the pressure switch would ease back, and the clock would restart with dire results to the recovery team. This is without doubt what had happened to the unfortunate trawler *Mastiff*.

The mine as a whole was an ingenious and devilish conception, and its complicated mechanism was most cleverly designed, meticulously made and beautifully balanced: indeed, in its way a mechanical masterpiece. Hitler had boasted that he would employ a secret weapon which would be impossible to counteract or to recover, and in this mine his scientists appeared to have produced the irrecoverable.

But however ingenious and carefully made weapons of war might be, they can be subject to mechanical defects in manufacture or in operation, like any other piece of complex machinery. Through a simple defect in the mechanism the two mines dropped in error in the shallow water over the Shoeburyness sand did not complete their intended circuits and failed to blow up. As a result the experts at the Portsmouth establishment were able to discover their secrets.

Soon more parachute mines which were dropped in shallow water failed to explode, and their recovery by other teams sent down from *Vernon* revealed that the Germans were now using a larger and more powerful version. This new mine had the same round-nosed cylindrical case and parachute arrangement as its predecessors, but it now measured over 8ft in length, with the same girth, and its main charge was now a hefty 700kg (1500lb) of a German explosive known as hexanite. To carry this heavier mine the parachute was now some 15ft across and attached to the mine by 28 plaited nylon cords. Mine recovery teams were to discover how much their girlfriends in clothing-starved Britain treasured these silklike ropes as cords for their dressing gowns. Unhappily, the nylon parachute panels were a revolting shade of green.

The most vital of the secrets that now came to light was exactly how the magnetic mechanism was made to respond. It was found that the Germans relied on a comparatively simple dip needle device, a magnetised rod held in position by a sensitive spring, which was activated by the sudden increase of south polarity when a ship passed close by. It was similar in principle to that used in the early British M-Sinkers in 1918 but subsequently discarded.

Now that the method used and the period over which the intensified magnetic field had to act before the mine detonated were known, the immunisation of ships and new effective sweeping techniques could be put in hand. Plans for such an eventuality had long been made, ships being passed over recording ranges at many ports and their individual 'magnetic signatures' noted. The unit used for measuring the intensity of a magnetic field is the *gauss* (introduced by Carl Frederick Gauss, 1777–1855, a German scientist) and it was found that the German mine was activated if it was subjected to a magnetic field of 50 milligauss (thousandths of a gauss) for a period lasting five seconds. The magnetic field of ships in European waters measured on average from 70 to 100 milligauss, so it was noted that these mines were on a comparatively coarse setting and needed to have a ship fairly close in order to respond to her magnetic field.

As we have already seen, the current required to counteract the ship's own magnetism would not only vary from one vessel to another according to each magnetic signature, but this could vary for each ship according to the kind of cargo she was carrying, such as railway rails, machinery, tanks, guns, or iron ore. Furthermore, the vessel's signature could become appreciably altered after making a long voyage.

At the highest priority, therefore, all kinds of ships were *degaussed* by having electric cables passed right round the hull and connected to the ship's own generators. When a current of the appropriate intensity was passed through these coils the magnetic field induced cancelled out the ship's, and she became immune to the German mine in its current form. This immunising current did not, it should be stressed, in any way assist the destruction of any mines.

The ships' officers in each case were supplied with a detailed chart of the current required for various parts of a voyage, allowing for any calculated variation in the magnetic field. In the early days the degaussing coils were lashed to the vessels' sides, but in this position they were highly vulnerable in docks, alongside other ships and in bad weather at sea. Soon, however, it was realised that the magnetic field generated would be just as effective if the coils were carried round inside the hull; and from then onward the degaussing cables were taken inside steel pipes below the main deck in much the same way as the vessel's own electric conduiting.

There were thousands of vessels to be treated, needing hundreds of miles of electric cable as well as trained personnel to carry out the work. But the manufacturers produced the materials and the teams of experts got to work, and within weeks naval vessels, merchant ships, passenger liners, all kinds of important vessels in fact had become immunised, and went to sea with a feeling of safety from the hidden menace.

Smaller ships such as trawlers, coasters, dredgers, harbour tugs, ferries and the like were treated by *flashing* or by *wiping*. In flashing, a cable was laid round the vessel at about deck level and a current of about 2500amps was passed through for a predetermined time, when the gear was removed and the ship remained magnetically immunised for a time that varied between two and five months according to circumstances. In wiping, a small vessel had a coil carrying a similar current passed up and down her sides, achieving much the same result.

The degaussing of ships, whatever method was used, proved on the whole highly successful, and during the early summer of 1940 the number of sinkings through magnetic mines dropped dramatically. But degaussing of ships was not the complete answer, especially in shallow water, and a practical method of sweeping and destroying the mines had to be found. This was the next vital step in the war against the German mine.

Chapter seven

SPECIAL PRESENTS FOR *VERNON*

A battle of wits was beginning. It was one thing to know that the mines the enemy were laying were magnetic, not of the sound operated nor pressure variety, but quite another to produce at short notice an effective type of sweep for them.

Knowing that the German mines acted on the sudden rise of a strong magnetic field, it seemed obvious that if you towed powerful magnets over the position the mines should be bound to take notice and detonate. Accordingly some sweeps were made up consisting of a long sweep wire which was towed, like the old A-Sweep, between two sweepers, but having wooden floats spaced at intervals to which were lashed iron bar magnets about 2ft 3in in length.

Called the Magnetic A-Sweep, it proved successful in destroying some mines off Portsmouth, but every time a mine was accounted for most of the sweep was destroyed and the bar magnets lost. It was also one of the most awkward and unseamanlike kinds of sweep ever invented: paying it out over the stern with a heavy clanking bar of iron to be manhandled over every few seconds, apart from having to haul it all in again at the end of the sweep, was the sort of job in that cold winter of 1939 to 1940 that the sailors who had to do it described in their lurid fashion.

Following this strong-magnet line of thought, the Admiralty had also put in hand what was called a mine destructor ship. A small steam collier of 2000 tons named *Borde* was selected for conversion as her engines were right aft, and this enabled an enormous electromagnet weighing some 450 tons to be installed, together with its generating plant, in the two holds. When energised the magnet was calculated to project a sufficiently powerful field to actuate mines at a considerable distance from the vessel.

Like the unpopular Magnetic A-Sweep the *Borde* did indeed

detonate a round dozen mines off Spithead and in the Thames Estuary, with such startling success that two other small colliers were converted for similar duties. But possibly the Germans had wind of the idea and purposely began to lay mines with a specially coarse setting, for one after the other the ships put up mines so close that they were badly damaged, while some of the crew suffered leg and back injuries. Less nerve-racking methods of sweeping were coming in, and the ships were transferred to other duties.

Another version of the floating magnet was the *skid*, basically a lighter or dumb barge on which was mounted a large diameter solenoid coil energised when required from a generating set aboard the tug. A number of mines were thus blown up in narrow channels and harbours around the coast, but of course, each time a mine was destroyed the skid went with it.

A later development of the same idea was the fitting of coils to stripped Wellington bombers. The coils were energised from a separate petrol generator set fitted inside the plane, and a number of mines were put up in shallow waters. But the planes had to fly so low – as little as 60ft above the surface – for the magnetic field to actuate a mine that they were liable to be brought down by the blast right under their tails, and Wellingtons at that date were highly valuable for other duties. They did, however, have considerable success on this duty later in the war when dealing with magnetic mines in the Suez Canal.

The age-old method of destroying mines by countermining, by lowering an explosive charge on to the mine and exploding it that way, was tried where a suspected magnetic mine was holding up shipping in a fairway. But with these ground mines it was at best a slow and uncertain antidote. A far more promising idea which was being developed, however, was to tow energised cables over the position so as to spread a strong magnetic field underneath.

All sorts of technical problems had to be overcome, but first results with this experimental sweep were highly encouraging. Called the Longitudinal Sweep, or L-Sweep, in its formative stages, its cables were made to stay on the surface by having wooden floats attached at close intervals. But the requirement was so urgent that the manufacturers met the Admiralty's demands handsomely by producing at short notice a cable encased in buoyant material which kept it afloat.

In its final form the sweep comprised two cables, one 160yds and the other 700yds in length, having an electrode at the outer end of each. They were energised with a heavy charge of some 3000amps supplied in short pulses by a diesel generating set aboard the sweeper. Working in pairs, the sweepers could tow the Double-L, or LL, at speeds of 12kts, which was better than most other sweeps had been, while covering a satisfactorily wide area of ground. The pulses were arranged to last from five to six seconds, calculated to be long enough to actuate the German mine as it then was, and were spaced about 60 seconds apart, according to the speed of the sweepers, so that the roughly rectangular areas swept would slightly overlap.

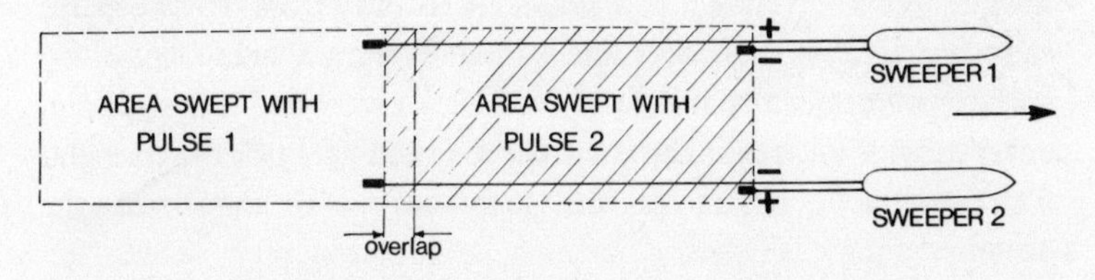

LL MAGNETIC SWEEP 1940–45 (not to scale)

Aboard the first LL-Sweepers the cables, which were not unduly heavy but measured some $3^1/_2$in over their covers, were coiled round the decks as they were hauled in. As they were about as awkward to flake down as a reluctant boa constrictor, the ratings handling them in wintry conditions used much the same pungent language as those on board the Magnetic A-Sweepers. But as more vessels became converted for the LL-Sweep, electrically operated drums, looking like giant cotton spools on the after decks, were fitted to take the cables. From then on paying out for a sweep, and hauling in afterwards, became mechanical routine, and no longer – well, less often – did the crew bemoan that they had ever joined the Navy.

And the LL-Sweep proved so effective in clearing the channels of the magnetic mine, and at the same time was so seamanlike in operation, that it remained the standard magnetic sweep throughout the years of the war.

It might not be out of place at this point to anticipate later developments of the LL-Sweep, which were to take place in the years following World War II and the Vietnam war.

Experiments in magnetic mine sweeping by the principal navies of the world resulted in many improvements being made to electrical sweeping methods, culminating in a type called the multiple magnetic loop sweep. This form of sweep claims many advantages over the old Double-L in that it can be worked by one ship only, in place of a pair of sweepers, while it can be arranged to cover the same width of channel as that normally swept by the Double-L.

Adapted to meet modern methods of mine clearance, this form of sweep utilises the same principle as the Oropesa float, which enabled the sweep wire to stretch in a broad curve away from the sweeper's stern quarter. The buoyant cable is made to form a large loop on the surface by means of a system of floats and a kite, and is designed to cover as wide an area as the older sweep towed by a pair of sweepers. The general principle of the loop sweep is shown in the accompanying drawing.

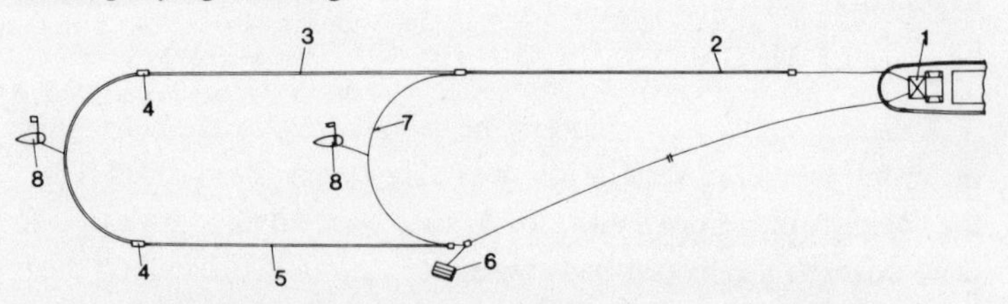

LATER VERSION OF MAGNETIC LOOP SWEEP
(not to scale)

1. *Connecting box on minesweeper*
2. *Married section, about 300 yards*
3. *Buoyant cable, about 150 yards*
4. *Buoyant electrode*
5. *Buoyant cable, about 110 yards*
6. *Kite*
7. *Catenary*
8. *Float*

Back to the days of 1940, then, it was clear that the falling off of sinkings of ships through magnetic mines in British waters was not lost on the Germans, and their natural mechanical ingenuity produced a number of obstructions designed to fox the work of the sweepers. Mines began to be laid with delayed arming devices which enabled the mine to lie inert on the bottom until such time, from two days to a week or more, that the period set ran its course and then the mine became 'live'. During this inert period normal sweeping methods would have no effect.

The mines recovered so far had reacted to south polarity. It was simple enough for the Germans to reverse this, and so render normal sweeping inoperative; but as soon as this move was suspected it was equally simple for all LL-Sweepers to be altered to give alternate north and south polarity pulses while sweeping, and so take care of both types of mine, or alternatively the two sweepers could change places for the next run.

Vernon by now had a number of RNVR officers trained in the techniques of rendering German magnetic mines safe under the direction of Lt Commander Ouvry RN, the officer in charge of the Enemy Mining Section. Specimens of every kind of mine the Germans had laid to date had been recovered and immunised by these RMS teams, and there was as yet no technicality concerning them that was not known by the British.

This prying into their secrets nettled the Germans into equipping their mines with anti-stripping devices, or booby traps. They knew that once the primer with its detonator had been carefully withdrawn from the main explosive charge, the mine was pronounced safe to transport to Portsmouth, and so to the mining shed for detailed stripping and examination. Now the heart of the mine, the compartment containing the arming clock and the magnetic mechanism itself swinging level in its gimbals, and surrounded by a ring of dry batteries with the various electric leads to the firing mechanism, was in the rear part of the mine case. This was closed by a circular rear door, a dished plate of the same alloy as the case, which was held on its watertight seal by 32 studs spaced closely around the rim; to open up this tightly packed compartment, like the engine room of a miniature submarine, it was routine to undo all the nuts, then to slide this plate off its studs and remove it.

One day a mine was being stripped in the mining shed, and when all the nuts were off, and the rear door was being eased back from its seating, there was a blinding flash and a roar. Four deaths resulted amongst the working party, and others were taken to hospital with bad burns.

This was the first of the anti-stripping booby traps, but as it was only a small charge that exploded, and fortunately not the main charge, the mechanism was still mainly intact. Examination revealed the reason why the Germans had been so careful to fit a booby trap to this mine: it was the first to be recovered that incorporated a new *clicker* device.

It was in itself a simple innovation which could be preset to actuate a selected number of times before the firing circuit was completed and the mine exploded. The Germans knew very well that in some convoys the leading ships were often the smallest and carrying the less urgent cargoes; the clicker mines were set, therefore, to allow several ships to pass before coming alive and bagging what was hoped to be a big one with a more valuable cargo. It was found that clicker mines might be set to operate up to sixteen times before becoming active, and once this was known the minesweepers' task was made that much more tiresome.

But in this cat-and-mouse business of mine warfare the British were not left behind. Indeed, the latest version of the British magnetic mine, which was laid by aircraft or by submarine, was a highly sophisticated design: it was both difficult and dangerous to sweep, its varied settings resulted in a number of German sweepers being sunk after they had pronounced the channels swept clear, and the ever-changing pattern of laying the various minefields was reported to be greatly lowering the sweepers' morale. The British hadn't been the first to use a magnetic mine all those years before for nothing, and their inventiveness was now proving a thorn in the Teutonic side.

Furthermore the uncanny ability of the mine recovery teams to obtain specimens of their most secret devices was undoubtedly an irritation to the methodical Germans, and the order went out that some way must be found to tempt as many of the highly trained British mine personnel as possible to examine a specially planted mine, which would be carefully booby trapped so as to exterminate them. This, it was reasoned, would be enough to put off the rest of the inquisitive *Vernon* people from meddling with any more German mines.

It was in August 1940 that a parachute mine was reported as having come down in a field a couple of miles beyond Portsdown Hill overlooking Portsmouth Harbour. The night had been fine and clear with an almost full moon, and if the pilot had been ordered to drop the mine in the harbour it seemed strange that he was at least six miles inland, when the whole stretch of the harbour would be clearly visible in the moonlight.

The Commander (Mining) – then Commander G Thistleton Smith – and Lt Commander Ouvry agreed that there was something odd about the whole incident as they raced to the spot in a *Vernon* car. When they reached the place near the little village of North Boarhunt, where the mine with its parachute still attached was being guarded by the local police and the Home Guard, their suspicions were fully aroused. There had been only a partial explosion, and pieces of the damaged shell were lying around, but the main charge looked to be still intact.

With great caution the routine of opening up parts of the mine was carried out from a safe distance, and there was no further explosion. But what was revealed confirmed the two officers' suspicions: this was no ordinary mine, for it *contained no magnetic mechanism*, but it was fitted with no fewer than three booby traps. There were two small charges wired up so as to explode if either the main fuse or the rear door plate was tampered with, and a third charge on any attempt to open up was intended to detonate the 1500lb main explosive . But one of these smaller charges had exploded on landing on the hard ground, and fortunately had disrupted the others' circuits.

The conviction that this mine was a deliberate 'plant' to try to catch mine disposal personnel was confirmed when a second parachute mine was reported near Piddlehinton, a small village in the Dorset countryside about ten miles inland from Portland Harbour. Two mines dropped on land near, but not on, the two principal naval bases on the South Coast were altogether too much of a coincidence.

Knowing what to expect in the way of unpleasant surprises from the North Boarhunt mine, Commander (M) decided that this second one, which showed no sign of any charge having blown, should be treated with even greater care as there was no need to hurry over the operation. Special X-ray equipment was accordingly taken to the spot, and revealed that, like the first, this mine appeared to have three separate charges as booby traps. So far, so good, but how to get at them without completing one of the electrical circuits?

The workshops were instructed to make an ingenious trepanning machine, basically a small engine which could be clamped on to the mine case and, driven under remote control by compressed air, would cut a round hole in the case large enough for a man to pass a hand. The engineers produced this in two days, and through the holes the little machine cut out the various electric leads were reached and the monster rendered inert, when it was transported on a lorry to join its twin already in the *Vernon* mining shed.

Until this incident the mine warfare had appeared as an entirely impersonal conflict between countries, but now those in the Enemy Mining Division of *Vernon*, officers, men and scientists alike, came to feel that they were being singled out for really personal treatment by the enemy. The directive must have come from the German High Command, possibly even from the *Führer* himself, to catch as many of those highly trained *Vernon* people as they possibly could. And had the planting of those two special mines been regarded with less suspicion, and attempts made to render them safe in the normal way, who knows how many of the specialist personnel in *Vernon* might have been wiped out?

Although reports of unexplained underwater explosions in the channels of the East Coast were not uncommon, as it was known that magnetic mines were occasionally given too fine a setting, and could be detonated prematurely through a variety of causes, the report received from a motor ship which had been mined and sunk off the Norfolk coast was puzzling. Evidence showed that it could not have been a contact mine nor a torpedo that hit her, there had been no air attack, while the suggestion that it might have been only another magnetic mine that had sunk her was ruled out, as it was established that her degaussing equipment was functioning correctly at the time.

It was while this ship's evidence was being examined that a report was received from the cruiser *Galatea* that while she was entering the River Humber at a speed of 20kts, a heavy underwater explosion occurred some 40yds ahead of her bow. Spray landed on her fo'c'sle head, but no damage was done to the hull. A week later, as the same ship was steaming through the swept channel of the Black Deep on her way towards the Thames, a similar phenomenon occurred. Again her speed was 20kts, but this time the explosion was reported as 50yds ahead of her bow.

Soon other reports of a similar nature began to come in, gradually filling in the squares of a regular pattern. Cruisers of different classes and destroyers steaming at speed described near-misses from mines which blew up off the bow or almost abeam, while there were now losses amongst degaussed merchant vessels, a tanker, four diesel coasters and a motor torpedo boat of Coastal Forces.

It was thought at first that the rate at which each vessel had been steaming was a contributory factor, the engines or possibly the propellers happening to give off a sustained note on a particular key. This certainly pointed to an acoustic, or sound-actuated, mine which might be circumvented if its sensitive sound range could be discovered, so that ships could be warned to avoid speeds that gave off such notes.

Then a large freighter was blown up when she was not even underway, but in the act of weighing her anchor. It seemed here that a combination of sustained sounds had triggered off the mine: the rhythmic clatter of her steam windlass, the rumble of the cable as it was hauled through the hawsepipe and into the cable locker below, perhaps even the subdued underwater noise as the heavy links rasped over the sea bed close to the waiting mine. All these sounds appeared to have activated the mine, but which one, and how?

That the Germans were laying acoustic mines in numbers was now acknowledged, but as with the early magnetic mines not much could be done to protect ships from them as yet. It had been observed that those ships which had put up a mine a little distance from them and survived had been doing between 18 and 25kts, while those which had been sunk were making their normal service speed of 10 to 12kts. Furthermore, small vessels, even slow coasters, which had noisy diesel engines appeared to be especially vulnerable.

Shipmasters were accordingly directed to reduce to half speed or even less while they were in mined areas, when it was thought their engines would be giving off the minimum of noise vibrations. The results of this precaution were satisfactory for a time, and there appeared to be fewer casualties amongst those ships whose captains had the patience to proceed at such reduced speed for hours on end.

Countermining by explosives, even the shaking effect of a 330lb depth charge, had proved ineffective, for the new mines did not seem to react to any sudden shock wave; they were evidently set to respond only to some form of prolonged sound, just as a ship makes when underway. If an excessive amount of *continuous* noise could be made under water, it was argued, than it should be possible to excite and detonate these mines at a reasonably safe distance.

But what was there in this modern world known to produce a fairly continuous deafening noise? A radio set, a brass band, a machine gun, a heavyweight tap dancer, a road drill, a riveter – ah, that was it! Something like a riveting hammer, but it would have to be amplified, and made to work under water.

While this line of experiment was being pursued with utmost despatch, from out of the lap of the gods (whom, in any case, the British believed to be firmly on their side) there fell a much desired specimen. It was in fact dropped by a *Luftwaffe* pilot into the mud of the Ogmore river in South Wales when he had been aiming at the channel leading to Cardiff Docks, but was dissuaded by the violence of the local anti-aircraft batteries. Once more *Vernon* rushed one of its RMS teams to recover it, and as soon as the dismembered pieces were laid in front of the technicians in the mining shed, they recognised it for what it was.

Indeed these scientists were already familiar with the principles of the sound-actuated mine, for they had been working on an acoustic design for the Admiralty long before the war. They were able therefore to discover very quickly that this enemy mine contained a relatively simple form of microphone connected to a reed which was tuned to a frequency of 240 cycles per second, in other words to about one note lower than Middle C on a piano keyboard.

This was broadly the range of sound, they knew from their previous experience, given off by many ships when steaming at their normal speed. It was patent, also, that any short shock wave such as that of an underwater explosion would have no effect whatever, for the mechanism required to be excited by a continuous sound until sufficient vibrations were built up to fire the detonator.

The riveting hammer type of noise was therefore deemed to be the answer, and the Kango was selected as the best available in large quantities. After various experiments had been carried out first with the hammer inside a flooded compartment in a trawler's bow, the final version emerged. In this the hammer gear was installed inside a steel box the shape of a short truncated cone with its wider end, which formed the diaphragm, a yard in diameter, facing forward. This box was at first fitted with a kite attachment which kept the box at a predetermined depth while being towed from the vessel's bow.

The Kango hammer proved so effective, detonating mines up to a mile away, that it became a standard fitting for merchant vessels as well as for many warships. In this form the noise box was mounted at the apex of a large A-bracket which was pivoted over a vessel's bow. When not in use it was raised, like a Thames barge's steeved bowsprit, and when needed for action it was lowered into the water ahead of the vessel's stem.

The threat of the acoustic mine seemed to have been averted for the time being, but the Germans were not to be thwarted so easily, and in their turn they introduced a variety of complications, such as delayed firing devices, combinations of acoustic arming and magnetic firing mechanisms, and more ingenious booby traps to dissuade the strippers. For the *Vernon* RMS teams even worse was to come.

Chapter eight

PUT THAT LIGHT OUT!

The failure of Germany's offensive with her primary secret weapon, the magnetic mine, which was expected to bring Great Britain to her knees in the first months of the war and ready to sue for peace, was one of Hitler's early disappointments. Indeed, it was the very first, for it had been confidently predicted by the German High Command that the British could not discover a practical counter-measure in time to prevent the majority of their ships at sea being sunk and their shipping lanes closed.

It was now galling to have to admit that the British had not only found a way to protect ships against the mine, but they had also developed a successful method of keeping their ship channels clear with an ingenious electric sweep. The secret magnetic mine had failed in its plan.

There was no need, however, for the ingenuity with which it had been manufactured to be lost, for it must be remembered that it had been designed as a dual purpose weapon: not only as a sea mine, but also to detonate as a heavy bomb when it fell on land. And there was a stock in hand of several hundreds already delivered from the factories. It may be apocryphal, but the story ran that in a cheek-puffing rage the portly *Reichsfeldmarschal* Hermann Goering bellowed 'Drop them on London! Let their other cities have a taste of them too.'

Accordingly in mid-September 1940 the Blitz on the capital began, first over the Docks where thousands of incendiaries were added to the parachute landmines, as they were called, and later indiscriminately over the whole of the Greater London area. Stories of quiet heroism shown by police and ARP wardens, by fire services and ambulance crews, by doctors and nurses, and by countless members of the general public have been recorded with pride in many forms. The people of London, experiencing for the first time the kind of holocaust that was to devastate so many cities before the war was to end, stood firm, hardened in their resolve to stick it out and save their homes if they could.

One small dividend gleaned from the method of functioning of the German magnetic mine was that a large number of them failed to explode. As they were officially sea mines and not bombs, HMS *Vernon* was called in to deal with them, so that life in the vicinity of each of these ominous bodies could be resumed as soon as possible, and people allowed to return to their homes. All available Rendering Mines Safe teams were accordingly hurried into the London area, found nightly accommodation in the basement of a large warehouse, and sent out each morning at daylight after the night's bombing to districts where unexploded mines were reported.

Some of these RNVR officers were greeted by puzzling situations which their own initiative had to solve. In one incident, for example, a young lieutenant found his 'baby', six feet of evil-looking cylinder hiding almost a ton of high explosive, half-buried in the backyard of a small terrace house with its tail leaning against the kitchen window. Its parachute was draped like a sombre shroud over the chimney pots on the roof.

The police had evacuated the whole street, and there seemed an uncanny silence over the scene as the *Vernon* officer took stock of the situation. He knew that the clock mechanism should have run off its 20 seconds or so on landing and detonated the mine, but for some reason it had failed to do so.

He pressed his ear to the side of the case. There was not a sound inside. If the clock had started and stopped after, say, ten or fifteen seconds, he reasoned, then any slight vibration like the unscrewing of the primer plate might start it running again, and if this happened it might be only a few seconds, and at the most about twenty, before the lot went up.

He had already taken careful note of any possible escape route, and seen on which side of the tiny garden the wooden fence was lowest, and the fences of the gardens beyond that. Again pressing his ear against the side of the monster, he took out the four-pin spanner made for the purpose from his box of non-magnetic tools, and carefully inserted it in the holes of the primer plate. So keyed up was he that, try as he would, he could not hold the spanner without shaking a little, and swore under his breath as the pins of the ring spanner rattled against the aluminium plate.

From inside the case came a muffled ticking. The young lieutenant dropped the spanner and cleared the fence in a flying leap, was across the neighbouring garden in a trice, but found the other fence higher and too much for him to leap over. But it was in poor shape, and as he charged it the rotten boards gave way under his weight. His body rolled over the top and he landed on his back in the next garden as the wooden fence fell on top of him.

As he did so it seemed as though the whole earth erupted. He had no recollection of the roar of the explosion, for the main blast must have passed over the shattered fence which protected him. But for minutes afterwards, it seemed, as he screwed himself up into a ball under the woodwork, pieces of earth and bricks and debris continued to thud all round him, but happily he was not hit.

Crawling out from under the wrecked fence he found he could walk, although his ears hurt and his eyes seemed full of grit. But all around him the scene had completely changed. Where the neat row of terrace houses had been standing was nothing but desolation beneath a black pall of dust and smoke. Twenty or more of peoples' cherished little homes had gone up in a flash.

Although he was alive, and thankful to find himself apparently uninjured, the thought of all those families now homeless through his own failure to deal with the mine began to prey on him, and at that moment, whatever others might think of the work he and others like him were doing, he felt saddened and far from being a hero. But tragic though it all was, this was routine work and he would have to pull himself together, for he had been warned that there was another unexploded mine to be dealt with half a mile away. ... He was one of the survivors and was later awarded the George Cross for this and other work during the London Blitz.

One of the more poignant memories the author retains of this tragic period, when he was put in charge of a naval diver and his lads to deal with mines which had fallen during the night into the Docks, was the scene which greeted them each morning as they threaded their way through the East End. Streets the diving party had driven through the previous day were now avenues of smoking rubble, and on occasion the police had to show them how their van could reach the part of Dockland where an unexploded mine was reported.

In places groups of people with drawn faces lined the pavements, but when they caught sight of the dark blue Navy truck hurrying through, their expressions lit up and they usually gave a cheer with Churchill's victory sign. Yet not one of them knew what horrors the following night would bring, what chance there might be of finding themselves and their families homeless, too, or no more.

During those grim days and nights of September 1940 there were many incidents which occurred to the *Vernon* teams as they went about their allotted work of clearing the wreckage of unwanted mines, and there were a few casualties amongst the RMS teams. Meanwhile the German High Command ordered the attacks to be carried beyond the outskirts of the capital. Landmines were now being dropped indiscriminately over towns and cities as widely scattered as Liverpool and Manchester, Coventry and Southampton, Chelmsford and Harwich, Birmingham, Sheffield, Hull and Glasgow.

The call for more and more trained RMS teams became urgent. Fledgling sublieutenants emerging from HMS *King Alfred*, the RNVR training school at Worthing, were rushed through the necessary instruction, issued with a standard kit of non-magnetic tools, and sent out with one of the experienced officers, for instruction in the front line.

The pioneer RMS teams who had been dealing with the earlier unexploded mines, and were later to receive the George Medal or George Cross according to their record, were given a few days' leave, and appointed afterwards to less trying duties. The author returned to the comparative peace of his Mine Recovery Flotilla command at Lowestoft, and was relieved once more to be able to take his little trawlers to sea off his familiar East Coast and deal with mines that were *under* water.

Continuing in their determination to use up their stocks of the magnetic mine on a variety of targets, the Germans staged a final blitz on the channels and harbour approaches of the East Coast, when sixty aircraft took part in a minelaying raid during one night in October. The results achieved by all these mines, however, were anything but spectacular, for most ships by now carried degaussing gear, while the well-equipped LL-Sweepers went out each day and dealt with them as a routine operation. Within days most of the infested channels were declared safe again. At sea the German mine blitz was evidently failing.

In December a parachute mine was recovered and on being opened up revealed a fresh development. This one was fitted with a magnetic mechanism having a bipolar unit, which would operate under either a north- or south-seeking magnetic field. This was patently designed to catch any ship which was either under- or over-degaussed, thereby presenting a field that would be strongly north- or south-seeking.

The advent of this variation on the magnetic mine theme was a logical outcome which had long been foreseen by the Admiralty, and indeed it occasioned surprise that the Germans had not invested in this type of mechanism much sooner. These bipolarity mines were accordingly taken care of by adjusting the gear of all the LL-Sweepers to reverse polarity with every pulse. Although this might appear only a simple solution to be immediately carried out, the delivery and fitting of new equipment during any war has its unavoidable problems and delays, and there were inevitably a few casualties amongst ships before the mines were swept and the channels cleared.

Soon after this there occurred two unexplained sinkings by mines in an area declared safe, and various methods were tried so as to excite and destroy any other mines of a new type which might have been laid. Then once again a *Luftwaffe* pilot came to the rescue when he dropped a mine in shallow water, whence it was recovered and duly transported to HMS *Vernon*. This specimen was found to have a new combined magnetic-accoustic unit with a delicate triggering device which enabled the mine to become armed by the sounds of an approaching ship, when the magnetic mechanism under the ship's field would fire the charge. This was set to such a sensitive degree that, once brought to life by the accoustic unit, the magnetic mechanism would detonate the charge under the influence of a very faint field indeed, such as might be found beneath a ship with slightly faulty degaussing equipment.

These conditions of approaching noise and a small magnetic field could, of course, be displayed equally by only a small vessel such as a trawler or a diesel tug, and the resulting loss to the war effort might be only trivial; but the Germans counted on catching big fish with this mine from time to time, and especially foxing the minesweepers as long as possible. Thanks to the recovered specimen, however, an effective sweeping technique was already in the pipeline, and the threat of the magnetic-accoustic mine was parried with the sweepers' customary determination.

Following on the intricate moves in this mine and countermine contest, the Admiralty received through certain channels which were trustworthy information that the Germans were about to introduce an entirely new type of mine, on which they were pinning great hopes, and which was described as being impossible to strip down and learn its secrets. It was said that the bickering and rivalry which had been such a refreshing feature through these months of war between Hitler's *Kriegsmarine* and his *Luftwaffe* had brought to a head the need for a sea mine which the *Luftwaffe* pilots would not – as the Admirals said – drop on land, so that the British could recover it and discover just what the Germans were up to.

The *Kriegsmarine* commanders thought, in any case, that minelaying should be entirely in the hands of the Navy, as their ships would be sure to lay the mines where they ought to be – under the water in the ship channels. They in fact lost no opportunity of accusing the *Luftwaffe* pilots of being careless and incompetent in their minelaying forays; on their side the air marshals declared that while they were expected to drop mines on parachutes accuracy of laying could never be certain – the things floated down too slowly.

What seems, then, to have emerged from these arguments was given in reports the code sign BM 1000. It did not need the expertise of a secret code breaker to infer that this could stand for *Bombe Minne*, and an inspired guess that it carried a 1000kg explosive charge. If this was so the Germans' new secret weapon was quite a formidable object. And if it was what the name implied then it was probably shaped more like a bomb, relied on no parachute, and would be dropped at some low altitude with all the accuracy of a normal aerial bomb. The possibilities of such a mine seemed endless, and the RMS people and civilian experts in *Vernon* awaited developments with quiet eagerness when news of the secret reached them, as it usually did.

To their grateful approval the *Luftwaffe* came up to expectations on a night raid in May 1941 and dropped such a mine on land outside Glasgow. It certainly was unlike any of the previous parachute mines, for it had a pointed nose like a bomb and a stubby domed tail from which plastic fins had disintegrated on striking the hard ground and lay in pieces nearby. The mine was painted a pale blue, measured roughly 6ft in length and 2ft 4in in diameter, and if there was any doubt that this was one of Hitler's new secret weapons it was soon dispelled, for stencilled on one end of the case as if to reassure the *Vernon* expectants was 'BM 1000'.

How this pastel shaded monster was carefully opened up by remotely controlled means, and all its parts rushed to the eager hands at Portsmouth, would make a nice story in itself. Let it suffice here to record that on examination it was found to contain a normal magnetic mechanism, except that it was elaborately mounted so as to withstand the shock of landing at high speed. As there was no parachute it was clearly dropped from a plane's bomb rack like any ordinary bomb, so that it could be sighted and laid with equal accuracy, while its rate of descent was some 550fs, compared with the 80fs downward drift of the parachute mines.

It was designed to detonate as a normal bomb on impact with hard ground, or after a 90-second delay if dropped in less than about 4 fathoms of water. If it sank into water greater than this depth a hydrostatic switch operated and set it 'alive' as a magnetic ground mine. Should a falling tide reduce the depth of water to less than 4 fathoms, or if the mine was transferred to shallow water or lifted for recovery, the hydrostatic switch moved again and the charge detonated.

Apart from this anti-recovery device, what had given the Germans such confidence that they had produced a mine that could never be taken apart for examination, even if the British did manage to recover a specimen, was revealed when the metal dome was removed from the tail of the mine. Two small circular windows in the casing, which became uncovered as the dome was drawn back, were found to contain photo-electric cells which were highly sensitive to light. They were wired to the detonator for the main charge, and if all had worked as intended the mine should have blown up.

When this ingenious anti-stripping arrangement was discovered, and the sighs of relief at its failure to work had died down, it was some encouragement to find the cause in a simple mistake in the wiring of the circuit: a result of carelessness in assembly, or perhaps the friendly fingers of a slave worker in the factory. By such remote chances can some of the most carefully designed and elaborate electrical mechanisms of warfare be made to fail to the enemy's advantage.

The discovery of this new type of bomb mine was treated with the greatest secrecy, but instructions on how to deal with it were immediately issued to the RMS teams at home. Warnings were stressed that the tail dome was to be removed only in complete darkness: no glimmer of light was to be allowed to fall on the photo-electric windows until after the mine had been defused.

It was not long after this that two enthusiastic Australian RMS lieutenants from *Vernon*, J S Mould and R H Syme, were together dealing with a BM 1000 which had half buried itself in a field near

Pembroke Dock in South Wales. It was a pitch black, almost sultry, night and they felt there was no need to erect any tent-like cover over the mine as there was no light anywhere near, for the general blackout in the district was complete.

The nuts holding the rear dome were soon off, the dome was slid back, and in the stygian darkness one of them could feel the little round windows of the photo-electric cells.

'It's okay, Cobber,' he said. 'Dome's off. Now for unscrewing the switch gear.'

It seems as if this remark caused the gods of the heavens to laugh their heads off, for the sky was split by a vivid flash of lightning followed by a sharp peal of thunder – and immediately another, and yet another, as the summer storm burst around them.

There was no time to run and take cover. Nowhere to run to, for they had not thought it necessary to dig a hide at a safe distance. The gods merely continued to roar with laughter, when just as suddenly as the storm had come, it died away.

The two friends breathed again and, without saying much, went on with their work of defusing the object in the dark. It was only later, after the mine had been examined and its mechanism tested in *Vernon* that it was discovered that the photo-electric cells required a light lasting a fraction of a second longer than a normal flash of lightning to excite the firing gear.

There were three Australians of the RANVR altogether amongst the *Vernon* RMS parties, Lt L V Goldsworthy in addition to Mould and Syme, and their combined record in this service was a highly creditable one. By the time the war was over Stuart Mould had been awarded the George Cross and George Medal, Hugh Syme the GC, GM and bar, and Goldsworthy the GC, GM and the DSC for special services. Australia was rightly proud of her sons, and after Mould and Syme died many of their personal effects were put on display in the War Museum at Canberra, while a portrait in oils of Goldsworthy hangs in pride of place on its walls.

Chapter nine

NICE OF THEM TO TELL US

It was not an occasion for surprise in high places when it was learnt that the Germans had begun to make concerted efforts to dam up the Suez Canal. Long-range bombers had already begun to make the long flight from air bases in Southern Italy and back to drop parachute mines of one kind or another in the narrow channels of the Canal.

Convoys of Allied ships had been regularly passing through, bringing to the European war zone urgently needed troops from Australia, New Zealand and the East, military stores and foodstuffs, oil in deep-laden tankers from the Persian Gulf, coal from India for the Allies' shipping and railways, and innumerable other cargoes destined for the British war effort. And like a thin blue line, as seen from the air against the sandy hills of the Sinai Peninsula, the Canal linked the Mediterranean at Port Said to the waters of the Gulf of Suez 100 miles to the south, leading thence to all the harbours of the African east coast and to the sea routes of the Far East.

The Canal follows a straight line to Lake Timsah, where the Canal Company's headquarters occupy the little lakeside town of Ismailia. Some 20 miles farther on the channel bends its way through the two Bitter Lakes, and thence in a straight line again past sandstone cliffs to Port Tewfik, the harbour for the bustling town of Suez. All the way the channel, which the French engineer Ferdinand de Lesseps had completed for opening in 1869, is so narrow that when two convoys of ships were approaching each other in opposite directions, all the ships of one convoy were made to tie up ('gare up' is the Canal term) alongside one bank to allow the other convoy to steam past at dead slow speed. The attraction ships passing in narrow waters have for one another would otherwise have caused innumerable sidelong collisions.

The Germans were well aware that if they could mine and sink one big freighter or tanker in almost any part of the channel, the whole Canal would be effectively blocked for many weeks, until the ship was removed or a new channel dredged out. Even the very presence of unexploded mines in the Canal would be effective in holding up all transits of ships until the minesweepers had got rid of them.

Early in 1941, following a moonlight raid by enemy aircraft, two ships were mined and were blocking the Canal. The first was a 4000-ton Greek freighter which was sunk with her back broken in the channel a couple of miles to the south of Ismailia. The second ship was fortunately somewhat smaller and had come to rest close in to the east bank about 12 miles north of Port Tewfik. In each case it appeared the mine had exploded just as the vessel's bow passed over it.

All ship movements for the time being were held up and salvage teams were at work on the wrecks, while the Canal Company's dredgers were starting to widen the channel opposite each casualty so that, when convoys were permitted to move again, the ships could edge their way past. Fortunately for Canal and shipping interests neither of the sunken vessels was a really large one, and it was not long before movements through the Canal could be resumed. But blocking this waterway completely was clearly a preliminary intention of the German High Command's designs on the occupation of the whole of Egypt.

Once the country of the Pharaohs was in Axis hands it would become an ideal base for the command of the whole of the Mediterranean; its value was immense, and the Germans meant to have it. Anticipating this move, the British had for some time been amassing their forces in the Canal Zone, sending out troops, guns, tanks, desert transport, even locomotives and rolling stock for military use on railways, while the RAF was building up its stocks of fighters and Wellington bombers, and all the equipment needed for the coming engagements, as rapidly as the planes could be spared from the hard pressed home front.

It was at this stage in the Middle East situation that HMS *Vernon* received orders from the Admiralty to send out to the Suez Canal forthwith an RMS officer with diving experience, together with a diving instructor who would train parties in the Zone, while the officer would initiate others in rendering mines safe techniques. It happened that the author had previously received a crash course in naval diving drill at Whale Island in Portsmouth Harbour, and was expected therefore to be able to take charge of diving parties. At that time frogman-type diving, carrying one's own air bottles, had not yet become a section added to the Navy's diving scene, and all the traditional deep sea diving gear – special diving boat with air pumps, regulation diving dress with lead boots, breast plate, brass helmet, air hose and communication line – was the order of the day.

Hauled back from leave in a countryside white under the snow with orders to report to the Admiralty forthwith, Griffiths found himself the next night joining a Wellington bomber's crew at an RAF base in Suffolk. With him was Leading Seaman N L Smith, an experienced diving instructor from Whale Island, who had also undergone instruction in dealing with enemy magnetic mines under water.

The 'Wimpey' was one of the bombers being stripped of all unnecessary equipment to have extra fuel tanks buttoned on so that they could stay in the air long enough to make the overnight hop to Malta. After being refuelled, and provided they were not damaged in the almost nightly bombing of the little island's air base, they took off the next night for the second long hop to Egypt and the RAF base in the Canal Zone.

At Navy House in Ismailia, where the Senior British Naval Officer Suez Canal Area (SBNOSCA), Vice-Admiral Sir James Pipon, had hoisted his flag, they learnt that diving gear and airpumps had been requisitioned from Alexandria and suitable boats assembled at Port Said. Commander D A Irvine RN had also only recently been sent out from England to take up his appointment as Area Mine Sweeping Officer (AMSO), and with the whole length of the Canal as his responsibility was faced with a formidable task. With admirable energy he soon organised sweeping operations with the limited material that was then available.

Apart from two small North Sea trawlers which could be put to work with ground sweeping gear, two of the Canal Company's steam sand hoppers – bulky vessels fitted for some reason with inboard-turning twin screws which made their steering somewhat unpredictable – had been converted with stern reels and electric generator sets to work the LL-Sweep. In most places the waterway was too narrow for these hoppers to work in pairs, but they achieved good results working singly or 'gared up' as static sweepers over suspected areas. In anticipation that the Germans would lay acoustic mines the trawlers and hoppers were fitted with hammerboxes on brackets over their bows.

With these vessels, together with one of the Wellingtons which the RAF had fitted out with an energised coil looking like an outsized halo, minesweeping was being carried out with commendable energy, and both types of sweeper were able to claim at least one kill. The Wellington also put up two magnetic mines near Kantara, but was so nearly brought down by the second eruption close under its tail that the authorities decided not to endanger any more aircraft or crews, except in dire emergency.

Leading Seaman Smith, soon promoted to Petty Officer, continued to organise diving parties, giving the necessary instruction so that systematic searches could be carried out over the bed of the Canal. Mines were located by this method, the diver attached a small countermining charge against the mine case and returned to the surface, and after the diving boat had moved to a safe distance the charge was fired and the mine exploded with an altogether satisfactory mushroom and plume of muddy water.

A signal had been received from HMS *Vernon*, however, that where possible a specimen mine must be secured and rendered safe, and if found to have any features of a new description, the appropriate parts were to be flown forthwith to Portsmouth for examination. There was always the probability that the Germans would be introducing something new and more unpleasant. While the minesweeping policy was pressed forward, therefore, in the task of clearing the Canal, Griffiths secured the services of two small drifter/trawlers which had managed to find their way out from home. These he had fitted with the non-magnetic trawls of the kind which had been in use by the *Vernon* mine recovery flotillas.

Following a similar technique of hauling any mine caught in the net to the shore, with the banks of the Canal so close it was a simple matter; on coming fast with a suspected heavy weight, the trawler moved about 400yds farther on, and put her bows in towards the bank. A heavy snatchblock was anchored in the ground some distance back from the water's edge, the trawl wire was passed through the block and brought back to the steam winch, and hauling began. In this way a specimen was hauled ashore, rendered inert and duly opened up with no dire results: but the second mine the little trawler captured exploded as she was hauling and shook everybody on board. It was possibly a timely reminder to treat all these objects with respect and care, for until the sting had been pulled out of the tail, as it were, a magnetic mine remained a menace to all around it.

One morning one of the divers located a parachute mine in the channel at the southern end of Lake Timsah, and reported that this one appeared to have some redistribution of the primer plate and hydrostatic switch pockets. Griffiths thought this could mean something different, and joined the party in the diving boat. The drill of getting the mine into a trawl and hauling ashore through a snatchblock on land could not be employed on this occasion as the shore of the Lake lay too far away through shallow muddy water, while it also happened that both drifter/trawlers were at that moment miles away working on another part of the Canal.

'I think I'll put on the dress,' said Griffiths, 'and go down and take a look at it myself.'

'I shouldn't do that if I was you, sir,' said the diver. 'My dress wouldn't fit you anyway.'

Lofty, as his mates called him, was a big man, and Griffiths took his point. A small man in an oversize diving dress with only limited training could be nothing but a menace on such a job.

'I'll take my tools down with me this next dip, sir,' the diver reassured him, 'and deal with it the ordinary way. You can see clear enough in this water, not like it was in London.'

Griffiths remembered that Lofty was one of the naval divers who had gone down and dealt with mines in Dockland during the first days of the London blitz – a thoroughly reliable man who knew what he would have to do. He accordingly sat and watched as the diver's mate screwed in the face glass, and then giving the traditional double pat on the top of the brass helmet, fed the airpipe and lifeline through ready hands as the diver disappeared beneath the surface, and the two ratings started to heave round the airpump handles.

Lofty was soon on the bottom, for the depth here was only some 5 fathoms, and the men stood waiting as the steady stream of bubbles from the valve in his helmet spread over the surface of the water. The minutes dragged by slowly.

One of the pump hands broke the silence.

'Hope old Lofty don't get his wires crossed down there,' he muttered. 'I wouldn't give much to be in his shoes right now.'

The diver's mate at the boat's gunwale shot him a scornful look. 'Look matey, if you *was* in his lead boots now, you'd ruddy well sink, wouldn't yer?'

The repartee broke the tenseness they each felt while they sat over the unexploded mine which was being 'dealt with' beneath them. There was a decided sense of relief when a few minutes later Lofty rose to the boat's ladder holding out his tool bag. In it with the non-magnetic tools was the familiar metal cylinder containing the priming charge, while in his other hand he held the small shining detonator.

'All safe now, sir,' he said when they had unscrewed his helmet glass. 'It looks pretty much like all the others after all. Shall I lay the countermining charge now, sir?'

When the mine had been destroyed and the boat motored back to the Base, at Navy House Griffiths was met by an anxious cypher officer.

'This signal was received this morning, sir,' he said handing over the pink slip, 'and we couldn't contact you.'

It was from the Admiralty, and merely said:

Diving on to unexploded mines with intention of rendering safe under water is to cease forthwith. Anti-stripping device is now being fitted to cause mine to detonate on primer chamber being flooded. All diving parties are to be informed.

Griffiths read the message twice before handing it back.

'Oh well,' he said as he thought of the mine they had been working on that morning, 'you live and learn – if you live.'

There was still the problem throughout the Canal Zone that during each night raid by aircraft mines could be dropped without being spotted, and proposals were being considered how this could be overcome. One suggestion that reached Navy House came from some high authority and recommended that tall posts should be erected along both banks of the Canal carrying netting made of jute. By day the posts would be lowered with the nets into the water, so as to allow ships to pass over them. At night the whole lot would be raised so as to cover the waterway. And the purpose of this vast cumbersome canopy of rope? After a raid, if holes appeared in the netting – there in the water beneath each one you will find a mine! This suggestion, which for some reason or other earned the name The Cunningham Network, was quietly pigeonholed, and may perhaps still be in the World War II archives.

In the ports and harbours at home, squads of minewatchers were on duty during any raid, and it was rare for a mine to come down on its parachute and not only be spotted, but its position so accurately pinpointed that it could be countermined by a charge lowered on to it. But with something like 100 miles of Canal from one end to the other the number of watchers that would be needed could just not be spared from any of the Services.

Following another suggestion, however, the Egyptian authorities were approached, and so pleased were they to be able to co-operate with the British Navy and help to keep their Canal operating that the Egyptian Army was instructed to supply the necessary men and labour. And excellent allies these cheerful little brown-skinned soldiers made for this duty. Chest-high slit trenches were dug along the East bank of the waterway about half a mile apart, where at each watching post they mounted a fixed board having a straight wooden pointer pivoted over a nail in its centre.

If during a night raid a parachute mine was seen to come down and splash into the water, the sentry lined it up with his pointer and, with a piece of chalk, drew a line along it on to the board. Two adjacent watch posts doing this gave a fair transit where the two chalk lines crossed, and it was found that this primitive spotting was surprisingly accurate. Not only did the Egyptian soldiers record every mine that was subsequently dropped in or near the Canal, but each one was found within a few yards of where the transit lines indicated – unless, of course, the mine had blown up soon after landing.

In the area of Lake Timsah and the Bitter Lakes, the Army supplied boats for the spotters, who could, of course, not use fixed transit boards. But it was found that the *Luftwaffe* pilots tended to avoid dropping their loads into the lakes, as the lie of the ship channel could not be clearly seen even in bright moonlight; and if any mines landed by mistake in the shallows the cunning British would be up to their tricks and recover one. At some posts along the bank, where they could be spared, Lewis guns were issued together with a gunner, and the presence of these, apart from the impressive noise they made whenever a German plane showed itself against the starry sky, gave wild satisfaction to the minewatching troops.

One moonlit night in May 1941 when there was a general raid not only along the Canal but on Port Said and Ismailia as well, minewatchers on the Canal at Toussoum, a village about 12 miles south of Ismailia, reported that a single plane had made a low altitude run along the waterway, and they had seen an object fall into the Canal with a very heavy splash. Unlike previous sightings this object had had no parachute, but had fallen just like a bomb, yet it had not detonated.

This looked like something entirely new, and the drifter/trawler *Landfall* was at once ordered to the spot. After the Admiralty signal there could be no more dives on to such an object, but it was thought if the vessel could get it into her trawl the bomb or mine might be hauled up the bank for examination, using the previous snatchblock and safe distance drill.

Landfall's skipper, Lt C Vine RNR, was as good as his word and managed to catch the object in his trawl during the first run over. But the trawl merely came fast, as the little vessel's engines were not powerful enough to drag the weight nearer the bank. The trawler moved ahead, paying out the trawl warp, until she was four hundred yards or so away, when her bow was put into the bank, and the wire was led through the anchored snatchblock on shore.

But even the steam winch stalled after it had hauled the object over the bottom until it reached the bank. There the steepness of the bank defeated it, and another approach had to be made. A tree-lined road ran along beside the Canal at this point; it was deserted now for it had been closed to traffic since the bomb had been sighted, and Griffiths thought of the ever-helpful sappers with their bomb disposal equipment.

In a short time a powerful six-wheeled truck with winding gear reached the scene in charge of an enthusiastic Captain Thompson, Royal Engineers, and some sappers, and accompanied by a young Canadian officer from the RN base, Lt G D Cook RCNVR. The bomb disposal truck's winding drum proved master of the job where the steam winch had failed ('Well,' said someone, 'I always said that old boat's winch couldn't pull the skin off a rice pudding!') and relentlessly – and at a safe distance – the trawl with the object inside it was hauled up the steep bank on to the road.

It was an entirely new type to those who examined it, a massive object with a bomb-type nose and small tail fins, light blue in colour all over, and weighing well over a ton. The question whether this was officially a bomb, and therefore to be dealt with by an Army disposal unit, or a mine which the Naval RMS chaps should handle, was cheerfully settled by the two Forces working on it together.

All safety bomb and mine rendering safe rules were strictly observed. After the plate covering the primer chamber was unscrewed a line was carefully attached and the plate drawn away from a safe distance, in case an anti-stripping charge was set to explode. The primer itself with its detonator was removed in the same way. The ring of bolts surrounding the rear dome which held the tail fins were unscrewed, and once more the party hauled the dome clear from their hide amongst the trees and waited.

No blinding flash and roar came to blast the palms, the relentless sun continued to beat down through the tree fronds marking the parched earth with patterns of light and shade, and the voices of the men by the winch truck sounded far away.

'It's a mine, all right, and your perks, Commander,' agreed Thompson as they once more examined the massive cylinder. 'But I don't quite understand what those two little windows are for. They're not inspection plates, for you can't see through them. But we've got our own patent trepanner aboard the truck. Would you like us to cut a hole in the case, say here where the mechanism compartment is likely to be?'

Griffiths and Cook jumped at the Captain's offer, and in a short while the little engine with its own supply of compressed air was clamped on the side of the mine, and while the group once more waited at a safe distance, it slowly drilled a neat round hole of hand's width in the casing. While they took it in turns to peer intently inside the cavity another RNVR mine expert arrived from Ismailia on a sporty motorcycle. Griffiths greeted the young lieutenant with a friendly grin.

'Hullo, Jamieson, what's this – a despatch rider's steed? Have you traded in your fleet camel?'

The younger man ignored the pleasantry, and his eyes opened wide as he looked first at the mine and then at the other three officers.

'Great heavens,' he exclaimed. 'It's a wonder you're still alive! This is one of Jerry's latest things in mines, a bomb mine, and those little windows are photo-electric cells. Let light into them when you move that rear dome and the whole thing should go up. It's the Germans' newest anti-recovery device. The signal about it came into Base only this morning, warning all mine officers. They've started dropping these things at home, and there's just been one recovered after a raid on Malta, so we're unfortunately not first in the field.'

'Then why didn't this one blow up on us,' Griffiths asked.

It seemed a good enough question at the time, but Jamieson shrugged his shoulders.

'No idea, sir, until we strip it down and have a close look.'

'Well,' the older man mused, 'it was nice of the Admiralty to tell us about it, even if their message did come a bit late.'

Because their mining campaign had so far failed in its aim to block the Suez Canal for more than a week or so at a time, the Germans began to increase their air attacks. Port Said was heavily bombed one night, and a week later Ismailia was subjected to a savage dive bombing attack evidently aimed at the naval base. In this first attack Navy House escaped with little damage and no casualities amongst its personnel, but the brunt of the screaming bombs fell on the native quarter of the little town, and left Egyptians mourning their dead and injured amidst the rubble of their pathetic little homes, bewildered by the cruelty of a war that hardly seemed their concern.

The German forces had already landed in Libya, and through the months of bitter strife to come, when Rommel and his panzer divisions came near to breaking into Egypt, the Suez Canal was a prime target for their bombers and minelaying aircraft. But the minesweeping forces had been augmented with more vessels and equipment, and the mine disposal parties multiplied. More RNVR officers were being trained in mine clearance and stripping techniques, and ports like Alexandria and Benghazi, Haifa and Beyrout now had their own mine disposal parties.

It was perhaps an ironical turn of history that, whilst the Germans and Italians with all their combined resources never succeeded in effecting a permanent blockage of the Suez Canal, it was the operators, the Egyptians themselves, who were later to dam up their waterway for years on end. But that was to happen during an entirely different conflict, and one which does not concern us here. With countermeasures against the enemy's minelaying policies throughout the Middle East now being well taken care of the original parties sent out from *Vernon* were recalled to Portsmouth.

There was evidently something in the wind on the home front, and after a short sick leave to recover from a bout of malaria Griffiths found himself appointed to the Explosives and Demolition branch of the *Vernon* mining section with plenty of urgent experimental work on hand. During the build-up of Allied forces in preparation for the invasion of Hitler's Europe, the long awaited Second Front, *Vernon* had received Admiralty orders to prepare suitable explosive charges in their hundreds for the purpose of sinking a number of vessels which were to be used as blockships.

It is generally assumed that mines are devised to damage ships' hulls from the outside, but it could be argued that an explosive charge which is placed against the inside of the hull can be regarded as a form of mine, like an internally attached limpet. Although there are no secrets to be guarded in methods adopted by any of the nations for sinking ships with scuttling charges, it is thought worth while to put on record here some of the techniques learnt during this operation.

In all 73 merchant vessels together with four warships, every one of them condemned for one reason or another, had been allocated for the purpose. They were given the code name 'Corncobs' and were destined to form breakwaters off the enemy coast to protect man-made harbours known as Mulberries: beyond this, no other information was as yet available to the demolition parties.

Originally the Director of Naval Construction, who was then at Bath, required that each ship should have four large rectangular holes, two on each side, cut in her plating at the waterline. The holes were to be covered with wooden tingles or patches, which were to be blown off when the time came by explosive charges fitted by *Vernon* personnel. The Admiralty stipulated that each ship, after the charges were blown, should sink upright, without heeling over, and settle on a shallow bottom in not more than ten minutes. Furthermore, the explosions must not otherwise damage the hull so that each ship could be salvaged when the operation was completed.

When it was realized how much valuable time would be taken up in drydocks while fitting all these vessels with the temporary patches, and just how vulnerable the ships when so fitted would be to even minor collision, *Vernon*'s proposal to use specially fitted explosives to punch a larger number of smaller holes through the hull plating and so obtain more even flooding, was finally adopted.

The condemned merchant ships were mostly coal burners and hence no longer welcome for convoy work, and ranged in size from an old-fashioned tramp of a type once described as 'built by the mile and cut off where you want 'em' – in this case some 350ft in length with four holds – to a cargo liner half as long again and with six holds and deep oil tanks. The majority were British, but others were of Norwegian, French, Belgian and Greek nationality, while the remainder were from the USA. Some of the last-named were Liberty ships, those all-welded steamers of some 7400 tons gross which were produced in their hundreds for the war effort by the remarkable industrial tycoon Henry Kaiser, who had never been in shipbuilding before the war.

Although none of these Liberty ships was over two years old they had each been condemned through some major defect such as faulty engines and boilers, rupture by a near miss from a mine, damage in a collision, or severe hull strain from Atlantic gales. The ships' hull plating varied in thickness according to the size and type of vessel from 0.4in in the smallest to 0.9in in the larger merchant vessels.

The four warships were to prove tough propositions with bottom plating in two of them averaging 1¼in in thickness. To determine how much explosive was going to be required to blow holes of a reasonably uniform size in the region of 10sq ft in the different types of ships, Griffiths with a small demolition team carried out a series of tests on various steel plates supplies from the Dockyard. Happily, two wrecks of cargo vessels of some 8000 tons each which had been mined and sunk in shallow water, one in the Thames, the other on the South Coast, were made available also for explosive trials. The *Vernon* party learnt many things while they blasted – or tried to blast – holes through the unfortunate ships' rusty sides.

At first steel baffle plates were tried welded close to the explosive charges, but these were found to be entirely inadequate to prevent the force of the expanding gases from escaping. In most cases the hull plating was merely bulged in the area of the explosion and a few dozen rivets blown out – not enough to flood a ship. Only a heap of a dozen or more sandbags with the explosive charge held in close contact with the hull proved sufficient to blow the required hole.

Amatol, a mixture of TNT and ammonium nitrate and available from trade sources in large quantities, was decided upon as the most effective explosive for rending tough steel plates. Standard cylindrical cases holding 10lb and 25lb charges were used as required.

All the ships were ballasted with pit slag or sand and gravel, and watertight bulkheads separating the holds, forepeak, coal bunkers and stern had large holes cut in them just above the level of the ballast to allow the water to flow freely forward and aft. The holes cut in the engine room and boiler room bulkheads were deliberately positioned 10 to 12ft higher so that these midship spaces would be the last to flood.

Of the warships, *Centurion* was a battleship of the 1910 era and scarred by years of ignominy as a target for gunnery practice in the Mediterranean, *Durban* a 'D' class cruiser of 1919 and *Sumatra* a cruiser of the Netherlands Navy and twenty-four years old, while the battleship *Courbet* represented the Free French Navy. These vessels each presented their own problems to an easy scuttling scheme because of the toughness of their bottom plating and the numerous watertight compartments in their double bottoms. Moreover, because of the size of the two battleships, both drawing some 30ft of water in their partially stripped and expendable condition, much larger holes would have to be blasted in their bottoms if these old warriors were to sink in the required ten or twelve minutes.

For the *Centurion* the Director of Naval Construction decreed that eight holes, four on each side, of an average size of 30 sq ft each would be required, and the groups of explosive charges were prepared accordingly. For *Courbet*, once the pride of the French Navy when launched in 1911, *Vernon* was left to devise the scuttling arrangements. She had been built five years after the appearance of Admiral Sir John (Jacky) Fisher's epoch-making all-big-gun

Dreadnought – she was in fact an improved Dreadnought, but now too old and outclassed in armaments as well as fuel consumption for modern war service.

Groping about with flashlamp and notebook in the gloom of watertight compartments and passages, through vast engine rooms and over the dozen cold and lifeless boilers, and deep down steel ladders into the noisome catacombs between the double bottoms, Griffiths and his party were conscious of the twenty thousand tons or more of steel and ironmongery surrounding them.

'How the heck,' he was compelled to ask, 'do you set about persuading a tough old battlewagon like this one to lie down and die on the bottom within a few minutes?'

'If you ask me, sir,' his Gunner (Torpedoes) said, 'it's like trying to fill up a big eggbox.'

Through the help of some friendly channels interior plans of *Courbet* were made available, and in collaboration with the Dockyard a scheme for flooding all the buoyant sections by torch-cutting holes in appropriate places – there seemed to be hundreds of them – was worked out. Against the 1¼in bottom plating 25lb Amatol charges were arranged in pairs in eight selected positions, each pair being tamped with up to twenty sandbags. In the event, as later examination showed, these charges proved adequate, for the holes they made were all roughly 5ft across, and the old warhorse settled upright on the bottom in under ten minutes.

In all the ships, merchant vessels and warships alike, the same arrangements for the firing gear were repeated. The primary scheme was by electric fuses connected to car-type batteries which were kept in a firing position right aft. As all the merchant vessels had been defensively equipped with a gun over the poop deck, a space beneath the now bare gun platform was chosen as it afforded good cover for the firing number in the event of air attack.

The electric cables were taken from here to all the explosive charges by way of the escape shaft along the propeller shaft tunnel to the engine room and thence to branch off to the pairs of charges. No elaborate firing key, traditional to all old naval demolition practice, was used: on the order to fire, the firing number merely took the two ends of the cable, removed the insulation sealing tape, and shorted them across the battery terminals. The resulting *crr-ump!* throughout the ship, as one of the firing numbers was afterwards to comment, 'was like a hearty kick up the pants.'

In the event of failure of the primary firing gear – through the electric cables being cut by bomb or shell splinters or any other cause – a secondary firing method was arranged, by which all the charges were interconnected by lines of instant-detonating fuse. These were led to a central position, usually under the bridge, and connected to a length of safety fuse giving a delay of about one minute, and a standard chemical igniter.

The whole of the fitting of the electric cables and detonating fuses, together with the arming of the many hundreds of explosive charges ready for the operation, was carried out entirely by *Vernon* parties in charge of RNVR lieutenants. It speaks well for the reliable way in which the work was done that, when the convoys of 'Corncobs' took up their allotted positions off the Normandy beaches following D-Day, all but one of the electrically actuated explosive charges detonated correctly. In the exception, the insulation covering of the cables had been eaten by rats, but the secondary detonating fuses did their work.

With one exception all the ships settled quietly on the bottom within the times specified. The rogue ship was *Alynbank*, a high sided ex-anti-aircraft ship and the first to be 'planted' as a mark vessel to the others. During the night the two tugs allocated to hold the 'Corncobs' in position while the charges were being blown had gone missing, and with a strong south-westerly blowing *Alynbank* heeled so much to the wind that the holes blown in her weather side were just clear of the water. During the time she took to settle on the bottom she slewed round and lay at right angles to the proper line. But her obstinacy was later to prove a blessing in disguise: she formed two conveniently sheltered anchorages on each side for an accumulation of small craft.

Sheltering the Mulberry harbours, the irregular lines of scuttled ships lay deserted, forlorn, rusting, yet immovable in the gales that blew that autumn and through most of the winter of 1944–45 while the Allied forces fought their way across Europe. In due time they were all raised without difficulty (none of the ships had been badly damaged by the scuttling charges) and salvaged for their considerable scrap value.

Examination of the effects of the explosives was found to add considerably to experience already gained in the technique of sinking blockships. Briefly it could be summarised as follows:

1 The explosive charges should be fitted to brackets welded in close contact with the hull plating, and each completely covered or tamped with from 12 to 20 sandbags.

2 Ballasting should be completed, and preferably levelled off to the required height, before the electric cables and detonating fuses are fitted.

3 Rats were found to gnaw through the soft coating of the detonating fuse, and through any exposed braided electric cable, threatening a complete failure in both systems. The fuse line and cables should be carried through conduit piping, or only armoured electric cable used throughout.

4 Both electric cable and detonating fuse should be secured every 3ft or less to frames, fore and aft stringers, angle irons, bulkheads etc to prevent swinging through the ship's movements at sea and becoming chafed through.

5 In fitting the lines of detonating fuse the parties employed should be instructed to avoid any sharp bends or kinks (eg taking the fuse tightly round frames or corners of bulkheads) where the detonation wave may fly off, and a failure result. It should be impressed that the fuse should follow as straight a line as practicable, with only easy curves like a miniature railway main line.

6 If enemy action by sea or air is to be expected, all electric cables and fuse lines should be given every available protection, such as along fore and aft stringers or as far below the waterline as practicable. The best protected route for the electric cables was found from each charge to the engine room, through the shaft tunnel and thence up the escape tunnel to the firing position.

7 It was also found advisable to provide some temporary fixing of the firing battery to the deck. When the charges were detonated in some of the ships the shock below decks caused the battery to leap up and scatter acid over the firing number.

As a lover of ships of all kinds, who would rather design a new ship than arrange to destroy one, Griffiths harboured very mixed feelings when the long lines of doomed vessels lay at anchor, stretching over ten miles of a Scottish loch and ready to sail for the Normandy coast. It was not without some emotion that he mentally waved goodbye to what he had come to think of as his 'poor old Corncobs'.

Chapter ten

PRESSURE ON ALL FRONTS

For many months the roads and railways in Southern England had been busy with military traffic. Early in 1944 vast accumulations of camouflaged tanks and guns, ammunition and stores, fuel tankers and road vehicles, assault troops, and all the paraphernalia needed to open the long awaited Second Front on the coast of Europe, were waiting for the signal to begin.

For these many months, too, the Germans had known of all these preparations, and with their air reconnaissance and secret agents they had watched the growth of the movements and made their own preparations to turn the invasion into a larger and more devastating Dunkirk. It seemed logical to assume that when the combined Allied forces did invade they would do so by the shortest sea route practicable across the Channel to some sheltered estuary or harbour where a landing could be attempted.

From the great build-up of landing craft the German aerial photographs had revealed along the coast between Folkestone and Margate, it seemed fairly certain that the landings would be somewhere in the Pas de Calais area. The Germans accordingly went about defending their occupied coastline in their accustomed methodical manner, and sowed enormous numbers of mines to protect all the estuaries and inlets and harbours along the whole of the coast of Holland, Belgium and Northern France.

Although the harbour of Cherbourg itself was well fortified and protected by minefields, it seemed highly unlikely – even if it was possible – for the Allied forces to switch their invasion plans to any part of the French coast farther west of Cherbourg. It was clearly the Germans' conviction, therefore, fostered by their reports of the state which the preparations in England had reached, that the invasion was due to take place at almost any time in the spring when the fortnightly high tides would serve, and in any case not later than the middle of May.

With this schedule decided upon the mines in all the protective minefields had accordingly been equipped with sterilisers, which were in effect self-flooding devices set to cause all the mines to sink to the bottom during the last few days of May. This was planned to make the work of clearing channels through the minefields necessary again at a later date when the attempted invasion had come and gone.

At the same time, the Germans were keeping with the greatest secrecy a stock of four thousand pressure mines in store at depots in Germany and in France. Because there was no known way of sweeping or recovering such mines once they were laid, their use was held solely as a last resort when all else had failed, and should the need arise the order was to come only from the *Führer* himself.

Now the time seemed to have arrived when every effort would be needed to prevent the threatened Allied landings on the Continent, and as there could be nothing to lose now if all areas that might be concerned were strewn with these unsweepable mines, like a last desperate throw of a gambler Hitler issued the fatal order.

It is not out of place to reflect that the original suggestion that a mine could be made to operate on the sudden surge and ebb of pressure in the water beneath a moving ship was attributed to *Oberleutnant* W Fett, of the German Naval Research Establishment, some thirty years before. And like so many new inventions, the principle of a pressure-operated mine had been thought of elsewhere, and had been investigated in Great Britain about the same time.

Between the wars the British had, in fact, developed what was called the oyster mine, and by 1939 had accumulated a small stock in hand. But like the Germans, even their experts could produce no safe sweeping or other countermeasures, and in view of this it was accepted that any policy to use them could only rebound in due course, and prove self-destructive. It is an axiom that once a belligerent embarks on using an irrevocable weapon such as the oyster mine, the enemy can follow suit, and the effect can return to the other's discomfort like a boomerang. This could be advanced as the reason why, during the whole of the 1939–45 war, poison gas or nerve gas, although held in stocks by all belligerents, was never let loose.

The pressure mine works on a principle that is familiar to all canal engineers, who know that as a ship moves over a shallow bottom the pressure wave being pushed ahead by her bow is followed by a sudden reduction in pressure in the water, and this in turn is replaced by an increase in pressure as the ship's stern passes overhead. It is this natural variation in water pressure that disturbs the canal or sea bed in shallow water, causes the familiar stirred up muddy wake, and in narrow waterways damages the banks if a ship passes at too high a speed; for the amount of pressure and general disturbance is directly dependent on the speed of the vessel and the depth of the water beneath her.

The oyster mine uses this appreciable variation of pressure beneath a moving ship to detonate the main charge. The mechanism is of the simplest type and consists of a chamber open to the sea which is divided into two parts by a thin metal diaphragm, the upper end of the chamber being covered by a flexible rubber bag or dome. When this bag is deflected by a pressure wave it causes the diaphragm to pulse, and this then makes an electrical contact and so fires the mine.

To ensure that the mine will not be inadvertently detonated by the relatively slow variations in pressure caused by a swell or surface waves, a small bypass valve connecting the two parts of the chamber keeps the pressure in both parts uniform, so that the diaphragm does not move. But this bypass cannot absorb the rapid change of pressure caused by a ship, and the firing circuit is then completed.

May 1944 came to an end, and still there was no armada. The spring had come and gone, and it looked as if the invasion scare was over, for the time being. Obediently, before the last day of the month, the thousands of mines along the occupied coasts became sterilised according to plan, and subsided harmlessly on to the sea bed.

Meanwhile the synthetic rubber bags of the pressure mines had come to the end of their term of usefulness, and with their diaphragms were scheduled for renewal in the factories. On *Reichsmarschal* Göring's own orders, therefore, all mines of this type in store were to be returned to Germany forthwith to have an improved form of bag and diaphragm fitted to them. From the various depots in Northern France the mines were accordingly loaded into rail trucks and sent on their way back to the Fatherland, and the very last consignment arrived from Le Mans at the Magdeburg arsenal on the evening of 4 June.

The following day Overlord, the Allies' code name for the greatest military invasion in the history of the world, was due to be launched. But so bad was the weather in this first week of June that the whole operation was held back for twenty-four hours. It was therefore just two days after the removal of all of Hitler's pressure mines, on D-Day, 6 June, that the first of the assault troops landed on the coast of Normandy, and the deliverance of Europe was begun.

So convincing had been the British display of dummy landing craft and military build-up along the coast of Kent and in the North Foreland area that the Germans were still satisfied that the landing in Normandy was only a feint, and that the main invasion attempt would be made somewhere between Calais and Boulogne. When it was realised that these landings on the beaches far to the westward of the Seine estuary and on the eastern side of the Cherbourg peninsula were indeed the main invasion of Europe, the Germans reacted with speed. As soon as further stocks of the pressure mines became available they were rushed by train for the Normandy area. But the intensive Allied bombing raids had so disrupted rail and road routes that it was ten days after the first assault landings (D + 10) before these mines could be laid in any quantity in the invasion area.

At Arromanches and other British beaches ships were issued with barrage balloons, and the presence of these blimps effectively reduced low altitude minelaying over the anchorages at night, although some of the German pilots got through and dropped their oysters amongst the shipping. Over the two beach-heads occupied by the American forces, at Omaha and Utah, on the other hand, less reliance was placed on barrage balloons. The *Luftwaffe* took advantage of the situation and dropped numbers of oysters with encouraging success. In the first few days they accounted for more than thirty different American craft, sunk or badly damaged by these mines, including four destroyers. These losses, which were quickly followed by others, added to an unexpectedly strong resistance put up by the German forces ashore, contributed to the decision to abandon the Omaha beach-head.

Meanwhile measures had to be taken to protect Allied ships from the threat of the oyster. Some areas were cleared by normal countermining methods, dropping explosive charges over mined areas and detonating the mines that way. But countermining this type of weapon was never one hundred per cent certain, and mined areas still had to be treated with suspicion.

Once more a specimen of one of Hitler's last secret weapons was obtained in a manner that had to be kept unheralded for some time, and its workings were duly studied. The amount and rate of change in water pressure required to operate the mine were discovered, and captains of ships were immediately notified of safety methods to be adopted when steaming into suspected areas. The faster a ship was going the greater would be the pressure pulse beneath her, so that the elements regulating the amount of disturbance were the size of the vessel, her draught or depth in the water, her speed, and the depth of water in which the mine was lying.

By reducing speed to some rate that might in the engine room be counted as 'dead slow' even a big freighter or a cruiser could pass over a dormant pressure mine in, say, 10 fathoms at 5kts with a degree of safety. Smaller vessels, such as a 2000-ton coaster or a destroyer, were given a speed of 8kts for the same depth, while for deeper water the safe speed allowed was a little higher. Very small craft like harbour tugs, MTBs, motor launches and so forth were not restricted, as it was not the noise of their engines that mattered – and some of them were always hideously noisy – but the small pressure waves they sent out. There appeared to be a certain amount of rule-of-thumb guesswork about such orders, but in the event they worked very well, and there were only a few casualties amongst ships from German pressure mines recorded.

While the invading forces advanced steadily eastwards like a swelling tide across the landscape of France, and first one port and then another fell into their hands, the problems of clearing the wreckage left by a retreating enemy multiplied daily. When American forces entered Cherbourg they were met with a scene of devastation, the fine harbour choked with sunken ships and mines, dock installations in ruins and cranes toppled into the water. And everywhere, in the countless number of mines themselves and amongst the wreckage ashore they found cunningly devised booby traps to blow up the unwary or the unobservant.

This was only a beginning to the months-long wrestle the liberating forces had to face to clear the deadly mines and killer traps in the trail of ruins that lay all the way through France and Belgium and Holland. At every one of the rivers and harbours and docks along the coast, at Le Havre, Dieppe, Boulogne, Calais, Dunkirk, Ostende, Zeebrugge, Breskens, Flushing and Antwerp – the names read like the stations of a cross-country stopping train – the task of clearing the wreckage was to have a familiar pattern. And the mine disposal parties, groups of specially trained officers and men from the American Navy and from HMS *Vernon*, met the challenge with patience and fortitude. Together they suffered casualties inevitable under conditions which could be said to reflect the motto of the hour: 'These are calculated risks which in the circumstances must be accepted.' And for some it was indeed their finest hour.

When the war in Europe came to an end on VE-Day, 8 May 1945, and people began to return slowly to the quieter conditions of peacetime, the task of clearing the Continental harbours of explosives and of minesweeping all the navigable channels had to continue. There were many thousands of mines to be swept and disposed of in areas all round the coasts, and still others left in awkward situations in docks and canals and locks by a resentful enemy. Many months were to elapse before it could be reported 'Channels are now clear'.

Meanwhile, although the war with Hitler had run its course, the war with Japan, the widespread fighting in the Pacific, was still being pursued with ever-increasing vigour. Under the combined efforts of the American, British and Australian air forces, the mining war was approaching its highest point so far in its history.

In the early stages, while enemy bases were systematically bombed by planes of the US Air Force, submarines hunted and destroyed Japanese naval units and merchant shipping. These latter operations were carried out with such vigour, however, that a shortage of torpedoes rapidly developed. At first the distances from the US bases to the widely scattered enemy targets were so great that minelaying by aircraft was scarcely practicable. It had long been shown, on the other hand, that surface craft were no longer suitable for minelaying in enemy waters, however fast or well armed the minelayers might be, since devastating air attacks could nullify all their efforts.

Since a temporary shortage of torpedoes had arisen, therefore, submarines were called upon for duties as minelayers, and they were supplied with 1100lb magnetic and acoustic mines which were of a suitable design to be ejected through the submarine's topedo tubes. Thus equipped, these undersea boats were able to carry out the lengthy operations necessary to reach enemy-held waters, and it is recorded that the first of these submarines to lay mines in Japanese waters were *Gar*, *Grenadier*, *Tambor*, *Tautog*, *Thresher* and *Whale*. Altogether, these six boats accounted for minefields as widely spaced as in the Gulf of Siam, the Gulf of Tonkin, off the coast of Indo-China, and in the Bungo-Suido approaches to the south-eastern entrance to Japan's own Inner Sea.

As the war progressed and the Japanese forces were driven out of first one Pacific stronghold and then another, air bases were established at progressively smaller flying distances from their targets. It became no longer necessary to rely on the submarine force to carry out such long sorties, and with only a few exceptions, for the remainder of the Far East war all minelaying was carried out by aircraft, planes of the RAF and Royal Australian Air Force also contributing their share.

The mines used by the US forces were mainly of the influence or non-contact variety, the magnetic and two versions of the acoustic type. One of these had a 1100lb charge for use in depths to 15 fathoms, while the larger carried a 1-ton charge for depths up to 25 fathoms, at which depth it was still effective against both submarines and surface ships. In addition, various methods of anti-sweeping and 'clicker' or ship-counting devices were in use, and all types were to prove highly successful in their effects.

The Japanese, on the other hand, never developed any sophisticated mines or mining techniques, and appear to have made little use of influence mines. Although they laid mines profusely throughout the Pacific war they seemed content to rely almost entirely on their traditional moored contact mines with chemical horns, a pattern which was little changed from the mines they had used with such effect during their war with Russia almost forty years before.

For some ship channels and harbour approaches they had the Type 92, a spherical contact mine with a 100kg (220lb) explosive charge and having eight chemical horns, while for deeper waters up to 500 fathoms they relied on the Type 93, a similar type of moored mine but with an increased charge. A third version of these mines was the Type 88, again a moored contact type, which could be laid by submarine. The Japanese also made some use of free-drifting contact mines which were cylindrical in shape, 4ft 1in long and 14½in in diameter with four electric switch horns at one end, which were fitted with tailfins for dropping by aircraft. And in the later stages of the war a mine called the *Camote* was introduced with a 500kg explosive charge. It was cylindrical in form with convex ends, 4ft 4in in length and 2ft in diameter, which was designed to be laid either from surface minelayers or with a parachute from aircraft.

The Japanese military policy indeed appears to have assumed a lightning war, following the example they had seen of Hitler's *blitzkrieg* in Europe, which would give them power over all the Pacific territories as well as Australia and New Zealand, which they had long coveted; and their plans had not included the necessity for elaborate mining operations. Again, their High Command appeared to have regarded the possibility of an enemy sowing extensive minefields in their own home waters as so remote that their naval forces had spent little time or effort in developing up-to-date sweeping methods which would be capable of dealing with every type of influence mine that could be expected.

By 1943 the Japanese were being slowly driven back into a crescent stretching roughly from the coast of Burma through Sumatra and the Java Sea, the Moluccas, New Guinea and the Marshalls Group, on a radius of some 3000 miles from the heart of Japan. This became known as the Outer Zone by the American forces, and from their bases in this zone it is recorded that 3231 aircraft, comprising USAF B-29s and B-24s, with the US Navy's TBF and PV-1 aircraft, aided by RAAF Catalinas and PBY-5s, made altogether 108 different sorties in which 9254 mines were dropped on Japanese target areas. Resulting from these operations 186 enemy ships were reported sunk and 154 more damaged by mines.

For the first time in history, also, minelaying was carried out from an aircraft carrier when planes took off in daylight from the US carriers *Bunker Hill*, *Hornet* and *Lexington* in April 1944, and raided Palau in the Caroline Islands. Flying through a barrage of ack-ack these planes dropped 78 mines and caused all channels to be closed to shipping for three weeks. So as to prevent ships from slipping out of harbour, mines were dropped in the entrance as soon as the raid began.

In August of the same year Superfortresses carried out the longest minelaying mission of the war. From their base in Ceylon they flew a 3800-mile round trip to the Japanese bases at Palembang and the Banka Strait on the north-east coast of Sumatra, where they dropped 16 mines. Whether so lengthy a mission with such a necessarily limited load of mines was justified may have been questioned, but the raid took the Japanese entirely by surprise and resulted in the sinking of three of their ships and damage to four others.

Apart from these results the moral effect on the enemy was felt to justify the expedition, and as one senior officer who was concerned with its planning was overheard to remark 'Well, we did it just for the hell of it.' It was certainly the unbeaten record for long distance mining.

Gradually, like the inexorable advance of molten lava down the side of a volcano, United States and British and Australian forces harried the Japanese bases, bombing installations, torpedoing ships, mining channels, driving the little men from one strongly held post to another, and forcing them to retract as the arc of fire steadily closed in. By March 1945 Allied forces occupied bases which stretched in a roughly formed crescent now within a radius of only 1500 miles from Tokyo as one Japanese-occupied island after another was overrun. No longer was it necessary to rely on submarines alone to approach the mainland of Japan and sow their mines, for the bases in this Inner Zone were well within flying distance, and a long-anticipated offensive of mine laying began.

From Guam and from islands in the Marianas Group B-29s carried out regular missions that resulted in no fewer than 12,000 mines of magnetic, acoustic and oyster type being dropped in the main ship channels of Japan. In the Tsushima Strait between the coast of Korea and the Japanese mainland – the very Strait where the Japanese Navy had destroyed the Russian fleet in the war of 1904–5 – as well as in the Shimonoseki narrows leading to Suo Nada, the Inner Sea, minefields were systematically laid.

The planes generally operated on moonless nights or during a spell of thick monsoon weather, when it would be most difficult for the enemy to spot and mark the positions in which the mines were dropped. They flew at altitudes between 5000 and 6000ft, and employed their radar for position pinpointing when visibility was too bad for visual siting; the PBY-5 Catalinas with their low flying speed operated at near sea level – 100ft or so altitude – to avoid detection by Japanese radar.

Altogether they created an almost complete blockade of Japan's southern industrial ports. The effect was becoming disastrous, for records obtained after the war showed that within the first four months the average of 800,000 deadweight tons of merchant shipping that normally entered these ports every month had fallen away to less than 180,000 tons, a drop of 77.5 per cent.

The faults in the Japanese war plans and their omission to allow for the possibility of intensive minelaying by their adversaries now began to show up. Their naval home forces were pitifully unprepared for an all-out attack by minelaying aircraft. When faced with more and more minefields on all their principal shipping routes as the Superfortresses flew in and delivered their loads, losses among the sweeper fleets became serious, and the morale of their crews fell to a low ebb. They were just unable to cope with the extent of the minelaying missions, nor could they deal with the variety and ingenuity of the mines themselves.

The losses in merchant ships in all the main channels and harbour approaches were alarming, and shipping carrying vital cargoes had to be diverted to the northern ports of Honshu. But this entailed transferring each cargo in all its forms to rail trucks, which then had to be hauled over a narrow gauge system – then 1.07m (3ft 6in) throughout Japan – which was slow and already heavily overloaded.

Shortage of materials of all kinds was becoming acute, not only in war supplies and equipment, but in food and all the necessities of life for a seething population. It was not without a grim sense of satisfaction, remembering what had happened at Pearl Harbour, that the Americans chose their code name for the whole minelaying offensive on Japan: 'Operation Starvation'.

Almost daily the scene became more desperate as the stranglehold took effect, and the Japanese nation began to reap the whirlwind they had chosen to sow nearly four years earlier. As the results of the mining campaign became more and more obvious, and Japanese shipping and naval forces found themselves confined to harbour, those in charge of American mining strategy advocated stepping up the number of operations and effectively sealing off the remainder of Japan's northern ports. They were convinced, and not without reason, that if they could get delivery of the necessary mine stores and intensify this campaign Japan would be brought to her knees, and within weeks, a few months at most, show herself ready to capitulate.

But as on so many other occasions during wars, the cry was 'Too little and too late' concerning the supply of mining material from the United States. In their frustration the advocates of intensive mining complained that the Chiefs of Staff at home could not appreciate the situation in the war zone (and how often has that kind of complaint been heard from the fighting line!) and by failing to deliver the fresh supplies of mines demanded, they were withholding this golden opportunity to starve out the enemy. They were not to know what was in the wind until, on the fateful day of 6 August 1945 a uranium bomb was dropped on the city of Hiroshima. Three days later this was followed by a plutonium bomb of even greater destructive power which devastated the great naval and industrial port of Nagasaki.

History records that the unbelievable horror of these two events forced the Emperor to come out into the open, and on the 14th of the same month Japan surrendered, while official records have shown that two further bombs of plutonium type were being prepared for dropping on other targets if the Japanese authorities had still held out.

The advocates of the atomic bomb were jubilant in claiming that victory had been won only through the use of their weapon. Their argument was that whatever had been the cost in civilian lives and suffering the war had ended abruptly, whilst if they had relied on a continuation of 'Operation Starvation' it could have continued for many more months, and in that time countless numbers of American and British prisoners would have continued to suffer and die in Japanese camps.

Nothing, then, short of invasion of the Japanese mainland, would have brought the war to an end, and as the fierce resistance met by the American forces when they overran Guadalcanal, Iwo Jima, Midway and all the Japanese-held islands had shown, the Japanese would prefer death to dishonour under defeat, and on their own homeland how much more dogged their resistance would have been. Surely the British people themselves would have resisted equally desperately had Hitler's forces invaded England in 1940.

Nevertheless, whatever the moral issues of the use of the two nuclear bombs might be, supporters of the less devastating all-out mining campaign were still convinced that, had 'Operation Starvation' been pursued relentlessly as they had planned, the Hiroshima and Nagasaki holocausts would not have been necessary. It is entirely conjectural, but the Pacific war against Japan might have been brought to an end, not by guns or rockets or bombs, nor even by invasion, but by the cumulative effects of the silent underwater menace, the mine.

It is not out of place at this point to consider what was achieved by the minelaying forces during the war against Japan. It is recorded that of the round total of 25,000 mines of all types laid in enemy waters by American, British and Dominion forces during this Pacific offensive, 21,389 were dropped by aircraft in 4323 separate missions, with a loss of 55 planes in all. Aircraft which took part in these missions included B-29s (Superfortresses), B-24s (Liberators), PBY-5s (Catalinas), TBFs (Avengers), and PV-1s (Venturas). In their operations from both the Outer and the Inner Zones the types of mine laid by the American forces comprised 12,053 magnetic, 4448 acoustic, 124 magnetic/acoustic, 2992 pressure/magnetic and 268 drifting contact. From all minelaying operations from the two zones records later obtained from the Japanese showed that 1075 of their ships, both merchant and naval vessels, were sunk or damaged, representing a total of 2,289,146 tons gross. This represents one ship sunk or damaged for every 24 mines laid.

From Japanese sources it was also learnt that at the height of the American mine offensive in 1944 and 1945, the average time a ship which had not been sunk but damaged by a mine required to spend under repair was from 70 days for small vessels to 100 days for large ships. And this was at a time when in all Japanese shipyards, drydocks and repair facilities were at a premium. Comparison of the average cost of the mines and the missions needed to lay them with the value of any single enemy ship thus put out of action bears out the claim that a well-organised mining offensive is highly effective for what it costs.

Against these figures, the Japanese themselves during the whole of their offensive in the Pacific distributed altogether some 70,000 mines throughout their captured islands, coasts and naval bases. With the exception of a small number of submarine-laid mines of the Type 88 spherical, and others of a large cylindrical type (the *Camote*), which were laid by surface ships or with parachutes from aircraft, almost all Japanese mines were of the spherical chemical horned Types 92 and 93.

Chapter eleven

SOMETHING OF THE FUTURE?

It has been shown in previous pages how the significance of the underwater mine grew from a tentative and haphazard device intended to sink any enemy ship that came into contact with it, to a form of strategic warfare which in certain circumstances could be the means of bringing a country to its knees. It might well be asked, then, where do we go from here?

The prospect seems to be full of possibilities. Indeed, many ingenious proposals in both the design and use of mining material have been put forward, and reflections on what some of these developments might amount to in any future wars can be offered only as personal opinions.

Whilst Bushnell's eighteenth century underwater attachment of an explosive device to the bottom of a vessel has turned full circle in the shape of the modern limpet mine, it is still the most direct and effective way of blowing a hole in a ship's hull; developed to home-in on a target like a miniature submarine, such a mine has interesting possibilities in rivers and harbours.

Against ships which are underway, however, the possibilities of the influence mine, whether magnetic, acoustic, pressure, or combinations of all three, are endless. It has been shown how a non-contact mine with a heavy charge detonated some depth beneath a ship is vastly more destructive than the hole blown in the hull by a contact mine. It has also been seen that such influence mines can be infinitely more difficult for an enemy to detect and sweep than the moored variety. It is more than likely, therefore, that influence mines will be developed to exert their destructive power in much greater depths of water than hitherto.

Refinements which are already with us include an electronic

programmer inside the mine which will respond only to ships of a certain type with a predetermined magnetic signature, or an individual sound track – the hydrophonic effects, in official language. This means that friendly vessels and neutral ships are able to proceed in safety, while there would be no reaction to electric or noise-making sweeping devices; but if an enemy ship approaches whose characteristic magnetic signature or sound track had already been fed into the computer and the mine sensor suitably programmed, the circuit will respond and the mine detonate. The introduction of that miniature marvel, the silicon chip, has made all this practicable.

It is not only in naval establishments that experiments in new mining techniques are carried out. Armaments firms also introduce new weapons and offer them under their own trade names. An example of this, from America, is called Captor. It is a self-propelling mine which could arguably be classified as a miniature torpedo. The explosive is housed inside a cylindrical case which enables the mine to be laid from a submarine's torpedo tubes, and the mine is moored by a short line to a sinker on the sea bed. A sensor in the head reacts to the propeller(s) of an approaching ship and releases the mine from its mooring. This starts a propeller driven by a battery motor and guided by the sensor the mine homes on the ship noises, and a proximity fuse detonates the charge when it is close enough to the hull.

Another commercially produced device for which many uses appear likely is known as Quickstrike. This is a neat 'do-it-yourself' form of conversion kit which enables a magnetic or acoustic sensor to be fitted inside the fuse pocket of a standard 500kg (1100lb) bomb. Thus converted to an underwater weapon the bomb can be dropped in the ordinary way by aircraft from considerable heights.

Speculating on future developments can reach the limits of credulity. Like Captor described above, mines can be moored in very deep water and released from their sinkers only on the approach of suitably programmed targets and detonated near the vessel by influence fuse. Mines of this type would present some problems for the minesweepers.

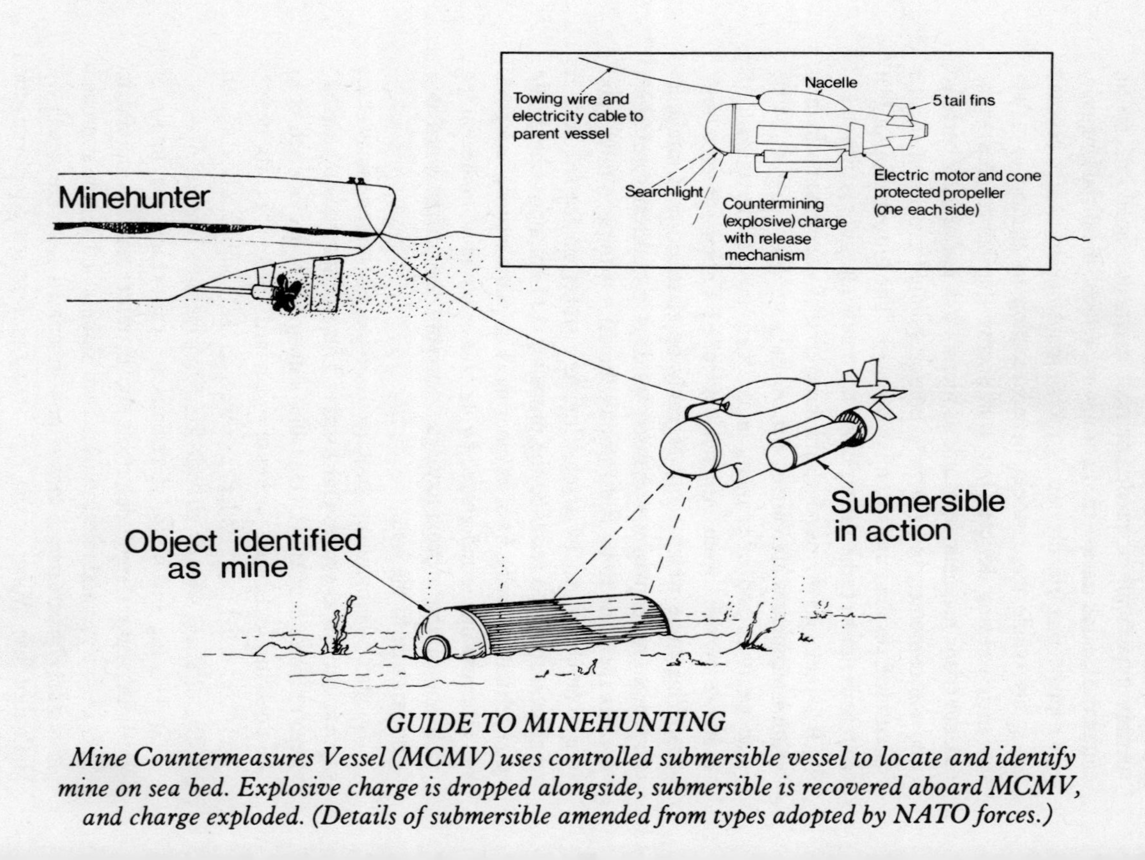

GUIDE TO MINEHUNTING

Mine Countermeasures Vessel (MCMV) uses controlled submersible vessel to locate and identify mine on sea bed. Explosive charge is dropped alongside, submersible is recovered aboard MCMV, and charge exploded. (Details of submersible amended from types adopted by NATO forces.)

Policies on mining techniques which have been carried out during successive wars have revealed the enormous potential in a well-conceived mining campaign based on what is known of the enemy's habits and thought processes: a successful mine barrage can be as much a psychological factor as mechanical. As an example of this, during the Korean War (1950–53) a mixed field of some 3500 moored contact and magnetic ground mines was laid by Communist sampans and junks to oppose the United Nations landing at Wonsan. These mines effectively delayed the landing for eight days while the United Nations forces gathered the necessary information and mustered enough minesweepers to clear the Wonsan approaches.

On another occasion, during the war in Vietnam (1965–73), the American forces in the spring of 1972 decided to mine the entrances to Haiphong and other North Vietnamese harbours as the most effective way of aiding the beleaguered South Vietnamese, since over 85 per cent of the war materials imported by the North Vietnamese was being delivered by sea. United States A-6 and A-7 carrier aircraft laid some 8000 mines of influence type, and the effects of this blockade were almost immediate: while 29 large ships were held up in Haiphong alone, no other vessels of any size either entered or left the harbours for the next ten months. During this period the North Vietnamese were reduced to bringing their much-needed war supplies from China overland by rail and truck routes that were under constant US air attack. In the conduct of this eight-year war it has clearly been shown how great a contribution this mining offensive made towards the peace agreement reached between the United States and North Vietnam in 1973.

* * * * *

The underwater mine, it can be said, came of age with the First World War, it matured during the Second World War, and now its future as one of the principal weapons of attrition and defence seems assured.

Some of its possibilities are outlined in the Prologue.

Appendices

Appendix 1

TYPES OF MINE

CONTACT

1 **Moored**	Buoyant, floating below surface, attached to mooring or sinker on sea bed
2 **Drifting**	Buoyant, in tidal or river currents, on float line at set depth beneath surface
3 **Creeping**	Buoyant, attached to length of wire or chain end of which trails on the sea or river bed
4 **Oscillating**	Free floating within a predetermined range of upper and lower depths below the surface
5 **Antenna**	Buoyant, moored, having length of copper wire antenna held up by float beneath the surface. Used against submarines or surface ships. Mine itself may have electric switch horns in addition for contact by submerged submarine.
6 **Hull attachment**	Commonly called the limpet, and applied by frogmen to ship's bottom. Fitted with time fuse
7 **Net mines**	Attached to anti-submarine steel ring nets which are held in position by lines of bottom sinkers and beneath the surface by lines of floats

NON-CONTACT (INFLUENCE)

1 **Magnetic**	Actuated by magnetic field of a ship in close proximity. Can be ground mine or moored
2 **Acoustic**	Actuated by underwater sounds made by a ship. Ground mine or moored
3 **Magnetic/ acoustic**	A combination of 1 and 2 in which the firing mechanism is armed by ship's magnetic field and actuated by the vessel's sounds. The process can be reversed
4 **Pressure**	Commonly called the oyster. Actuated by variation of pressure in the water beneath a moving ship. Ground mine
5 **Saddle charge**	Laid on bottom by midget submarine beneath target ship, with time fuse
6 **Controlled mines**	Usually in groups connected to an electric loop and fired from observation posts on shore. Used for protection of approaches to harbours, narrow channels etc. Moored or ground mines. Methods of firing may be electronically selective
7 **Self-propelled**	Moored, buoyant, and released by passing ship's propeller sounds or magnetic field, homes-in on ship with battery driven propeller, and detonates under ship's hull by proximity fuse

Appendix 2

METHODS OF FIRING MINES

1. Tumbler weight with friction fuse (obsolete)
2. Lead horn with chemical igniter (obsolete)
3. Herz horn with chemicals to form firing battery
4. Switch horn with electric contacts to firing mechanism
5. Antenna with copper plate, contact between antenna and steel hull completes a battery circuit
6. Magnetic, vitalised by various types of magnetic field produced by steel ship in proximity
7. Acoustic, excited by underwater ship-making sounds
8. Magnetic/acoustic, armed by ship sounds, fired by magnetic influence, or vice versa
9. Pressure, excited by variations in water beneath moving ship
10. Combinations of 7, 8 and 9 above

Appendix 3

MINE COUNTERMEASURE METHODS

MOORED MINES

1 Ground chain or sweep wire towed between two vessels (obsolete)
2 Sweep wire and cutters with Oropesa float towed by one vessel

MAGNETIC MINES

1 Magnetised coil on sled or catamaran towed by one vessel (obsolete)
2 Double-L electric pulse sweep, with two separate buoyant cables towed by pair of sweepers (obsolete)

ACOUSTIC MINES

1 Underwater electric hammerbox attached to vessel's bow (obsolete)
2 Mechanical underwater noise-maker towed astern (obsolete)
3 Acoustic generator and monitor, supported at controlled depth by floats, with control console aboard towing vessel

PRESSURE MINES AND VARIATIONS OF ABOVE

1 Powered catamaran, remotely controlled, simulating selected ships' pressure signatures (being developed)
2 Mine destructor submersible vehicle, unmanned, radio controlled on special wire, with battery drive to variable pitch propellers. Fitted with searchlight for TV camera to send recording pictures through wire to monitors on board parent vessel. Gyro compass in vehicle constantly records heading of

'fish' during operations. Expendable ballast rope trails on sea bed to keep vehicle at correct distance. Explosive charge is laid from vehicle by controller on board parent vessel, vehicle is then recovered, charge is blown, destroying mine

Appendix 4

THE HAGUE CONVENTION

Rules affecting the manner in which minefields might be laid in wartime, and their possible danger to neutral shipping, were drawn up at the Hague Convention of 1907, and were agreed by most – but not all – of the world's naval powers. The overriding principle on which these rules were based was the one that held that the seas and oceans of the world should be kept free for the peaceful use of shipping of all nations not engaged in warfare. The principal clauses can be summarised as follows:

1 This prohibited the use at sea of mines of the free floating type (described as 'unanchored automatic contact mines') unless they were arranged to become harmless within *one hour* of laying. This envisaged a ploy by minor naval forces whose ships might be chased by units of a superior fleet, and might drop free floating mines overboard in their wake. The chances of thus sinking a pursuing warship would be so remote, however, and the method so wasteful of mines, that few navies have ever made use of this type of mine. But the organisers of the Hague Convention felt it their duty to prevent what could be an indiscriminate menace to shipping.

2 This Clause forbade the laying of 'anchored automatic contact mines' which did not become harmless on breaking loose from their moorings. This insistence, that when the upward pull of a buoyant mine on its mooring rope is released, through the rope parting or the mine breaking adrift, a spring device must be incorporated to open the firing contact and so render the mine safe, has been followed in the design of moored mines ever since. But in practice, after a mine has been in the sea for a week or two, perhaps even months, all the parts become corroded, and the safety trigger device gets seized up and fails to operate.

When sweepers cut the mooring wires and the mines bob to the surface, or the mines break adrift and are carried on to a nearby beach, it is as well to treat them as still very much alive, and to deal with them by rifle fire or to render them safe by dismantling the firing mechanism.

3 Belligerents and neutrals who protect their own harbour approaches with minefields must issue Notices to Mariners giving the positions of minefields which might be dangerous to neutral shipping. In the two great wars the British and the Americans and their allies were careful always to adhere to this rule, and on more than one occasion, because of this law-abiding approach, the enemy obtained the appropriate information and made free use of the safe channels. On the other hand, it was possible to lay a few small minefields scattered over a wide area, while announcing the whole area as mined, and rely on a piece of bluff. Enemy ships would keep clear while friendly shipping could follow the secret channels with impunity. It was not unlike the yard notice 'Guard Dogs Loose' when the only menace is a friendly spaniel, but it often worked.

4 This fourth clause made it the duty of all belligerents after hostilities had ceased to clear all the mines they had sown. This was reasonable enough, but the disposal of mines which had been designed deliberately to make it difficult for the enemy to sweep – or the use of virtually unsweepable mines like the oyster or pressure mine – rebounded on the peacetime mine clearance forces.

INDEX